SpringerBriefs in Applied Sciences and Technology

Series editor

Janusz Kacprzyk, Polish Academy of Sciences, Systems Research Institute,
Warsaw, Poland

SpringerBriefs present concise summaries of cutting-edge research and practical applications across a wide spectrum of fields. Featuring compact volumes of 50 to 125 pages, the series covers a range of content from professional to academic. Typical publications can be:

- A timely report of state-of-the art methods
- An introduction to or a manual for the application of mathematical or computer techniques
- A bridge between new research results, as published in journal articles
- A snapshot of a hot or emerging topic
- An in-depth case study
- A presentation of core concepts that students must understand in order to make independent contributions

SpringerBriefs are characterized by fast, global electronic dissemination, standard publishing contracts, standardized manuscript preparation and formatting guidelines, and expedited production schedules.

On the one hand, **SpringerBriefs in Applied Sciences and Technology** are devoted to the publication of fundamentals and applications within the different classical engineering disciplines as well as in interdisciplinary fields that recently emerged between these areas. On the other hand, as the boundary separating fundamental research and applied technology is more and more dissolving, this series is particularly open to trans-disciplinary topics between fundamental science and engineering.

Indexed by EI-Compendex and Springerlink.

More information about this series at http://www.springer.com/series/8884

Csaba Visy

In situ Combined Electrochemical Techniques for Conducting Polymers

Csaba Visy
Department of Physical Chemistry
   and Materials Science
University of Szeged
Szeged
Hungary

ISSN 2191-530X   ISSN 2191-5318   (electronic)
SpringerBriefs in Applied Sciences and Technology
ISBN 978-3-319-53513-5   ISBN 978-3-319-53515-9   (eBook)
DOI 10.1007/978-3-319-53515-9

Library of Congress Control Number: 2017931348

Printed on acid-free paper

This Springer imprint is published by Springer Nature
The registered company is Springer International Publishing AG
The registered company address is: Gewerbestrasse 11, 6330 Cham, Switzerland

Preface

Over the past three and a half decades since the discovery of electronically conducting polymers, the large-scale potential applications have been evidenced, and many of them have been realized or available on the market. To improve the functionality of such polymer-based devices, the multifaceted cognition of these materials is indispensable. Elucidation of their redox switching between nonconducting and conducting states is of prime importance, requiring overall knowledge on the details of the complex process, triggered by an electrochemical transformation.

In situ electrochemical techniques are widely applied to study this phenomenon, because they deliver extra information about the processes from an additional aspect. Thus, color (or other spectral) modifications, mass, structural, volume, and conductivity changes—occurring in parallel with the electrochemical process—provide the complete knowledge by enlightening secondary effects of the electrochemical perturbation. The book is dedicated to the advantages of such combinations enabling to correlate various additive features, provoked by the electrochemical change as the common background. There is a fairly large number of books and book chapters available dealing with either conducting polymers or in situ electrochemical methods. However, only a few are dedicated to in situ methods applied directly to this family of redox active modified electrode layers.

The aim of this book is to link the most prominent properties of conducting polymers with the respective (combined) analytical methods which enable the in-depth understanding of such behaviors. This approach is indeed important, because it brings scientists closer to the rational design of new polymeric and hybrid materials for various applications. Moreover, recent technical developments have led to combination of two in situ analytical methods to enhance the understanding of redox behavior of conducting polymers by furnishing information on the self-same polymer film from several aspects at the same time by employing two in situ electrochemical techniques in a hyphenated mode. Despite spectacular developments in the last decade, there is no book or even a book chapter on the hyphenated in situ techniques.

Beyond the researchers involved in electrochemistry and conducting polymers, the materials science and especially materials chemistry/physics communities will also benefit from this book. Analytical chemists may also be interested; however, our primary goal is not to analyze the technical details of the presented methods, but to demonstrate their relevance to better understand conducting polymers (CPs). Importantly, this work extends further than CPs, since, to some extent, the presented methods can be implemented for various other electroactive materials as well, contributing to the work of anybody working with chemically modified electrodes.

This book is primarily aimed at an academic audience, but will also be useful for a wide range of professionals at different career stages, ranging from students to experienced researchers. As practicing educators, our aim is to provide an overview of the field. This approach makes the book useful to a wide range of people, from masters students to any scientist who wants to investigate and understand the electrochemical and coupled properties of electroactive materials.

Szeged, Hungary   Csaba Visy

Contents

Abstract

In this book, the advantages of in situ combined electrochemical techniques applied to conducting polymers, which enables the correlation of various additive features provoked by the electrochemical process, are summarized. The presentation of coupled phenomena during the process is followed by a brief overview of in situ combined electrochemical techniques. These methods deliver extra information about the processes via additional aspects such as color or other spectral modifications, mass, structure, volume and conductivity changes, which occur in parallel with either the electropolymerization process or the redox transformation. Their relevance is demonstrated by a better understanding of conducting polymers through a selection of applications, where problems and answers attempted by the in situ combined methods are presented. Advantages of recent technical developments—which have led to the combination of two in situ methods, furnishing information from several aspects at the same time on the self-same polymer film by employing two in situ electrochemical techniques in a hyphenated mode—are also discussed.

Keywords Electrochemistry · Conducting polymer · Polymerization · Redox transformation · Coupled phenomena · In situ techniques · Combined methods · Hyphenated mode

Chapter 1
Introduction into the Field of Conducting Polymers

In the everyday life, one meets large scale of various polymers, the common definition thereof being macromolecules consisting of repeating units, called monomers. Their several representatives can be discovered widely in our surrounding, since they are building stones in the nature, constituting living organs: polymers of hydrocarbons, such as sugars, starch, and cellulose on the one hand as well as polypeptides on the other hand, these are essential compounds for the life on our earth.

The other class of macromolecules comprises artificial materials, the big discoveries of the twentieth century, called plastics. Contemporarily, life is unimaginable without a good couple of their representatives from polyvinylchloride to Teflon, from bakelite to polystyrene, from nylon to polyurethane. The common property of these materials is not only their macromolecular structure, constituted from small chemically distinct units, but also the fact that they are well-known excellent electric isolators.

During the last three and a half decades, a family of macromolecules, exhibiting a surprisingly new behavior, has been uncovered. This group of polymers may behave as isolators, however they are able to conduct electricity under well-defined conditions.

This ability can be tuned by the oxidation state of the given polymer, leading to some representatives possessing a conductivity characteristic for metals. Even already at around the end of the 1970s, the conductivity range of the most fundamental polymers has been charted that were found to be regulated in a range of 6–8 orders of magnitude.

The background insuring opportunity for making them conductive is the conjugation along the polymer chain. In their neutral form, they are isolators or semiconductors (depending on the width of the band gap). Upon oxidation, the removal of an electron from the valence band creates charge carriers, a positive hole and an unpaired electron. *Delocalization* to conjugating neighbors creates a new type segment along the chain of the macromolecule, resulting new excitation band(s) at

© The Author(s) 2017
C. Visy, *In situ Combined Electrochemical Techniques for Conducting Polymers*,
SpringerBriefs in Applied Sciences and Technology,
DOI 10.1007/978-3-319-53515-9_1

energy levels above the valence but below the conduction band. Consequently, they appear within the gap.

Basically two types of conjugated polymers may exist: a form with degenerate ground state—the lonely representative being polyacetylene—and the rest of such macromolecules, possessing a non-degenerate ground state. They have aromatic or heteroaromatic ring system in which the position of the theoretically alternating single and double bonds is not energetically equivalent. The mesomer structures here are evidently different, representing a transformation between aromatic and quinoidal forms.

In the general context, the term "conducting polymer" comprises all kinds of macromolecular systems, which make possible electricity to flow. In this sense, the above-introduced CPs can be considered as intrinsically conducting polymers, since the polymer chain itself is electrochemically active in the electron transfer. This active participation is possible, because the charge carriers, delocalized along the chain, extending the unpaired electron and the yet uncompensated (see later) positive charge to some neighboring monomer units (from now on "segments") are mobile in an electric field, leading to the macroscopic conductance and to current flow via the charging/discharging of the given segment.

Another type of CPs can be characterized as a redox polymer, in which distinctly distributed unmovable, consequently localized redox centers are able to take up and transmit electrons forced to move by the electric field.

Again in an other case, the organic layer "is conducting" due only to its permeability for ions, although being and remaining uncharged, and just ion traffic occurs through the channels of the otherwise insulating macromolecule. In practice —in contrast to the description of the idealized intrinsically conducting polymers— these three sorts of behavior do appear mixed, since conjugation along a chain is not perfect and infinite, requiring electron hoping at discontinuities (as in redox polymers), moreover also electroneutrality requires the presence and traffic of charge compensating ions, a contribution to the conductance like in ionically conducting membranes.

One has to be aware that charge transfer and consequent transfer phenomena between the electroactive film and the adjacent electrolyte are not restricted just to CPs, so the in situ methods which are in the focus of this book can be applied also in studies on the wide spectrum of various redox active layers.

Chapter 2
Basic Features During the Redox Transformation of CPs and the Coupled Phenomena

2.1 Electrochemical Perturbation, Leading to Phenomenological Changes

As it follows from the previous chapter, electrochemistry is the basic tool to control both the synthesis and especially the redox switching between the states: the isolator/conductor behavior can be tuned through the level of oxidation hence it regulates the number of the charge carriers. Generally, the electrochemical transformation of a CP between its nonconducting and conducting forms can be easily achieved by performing cyclic voltammetry in an adequately chosen potential range. Under optimal conditions (the details are beyond the content aimed in this book), the transformation is chemically reversible, meaning that the total charge during the cycle is zero: all the material oxidized during the positive scan is reduced on the reversed section. However, the voltammogram is to certify the lack of the electrochemical reversibility: the overall processes are more complicated than it follows from the simplified description in the introduction. During the oxidation half-cycle one broad current peak can be generally seen, but on the reduction side the overlapping of two peaks are revealed almost in all cases. It is also typical that at the anodic turnpoint the anodic current tends to reach a minimum nonzero level.

2.2 Color Change

The above-performed electrochemical transformation results in dramatic color changes: during the oxidation the excitation of the film becomes possible by lower energies, leading to a bathochromic transformation. These color changes typical for the actual family of CPs are perceptible to the naked eye, e.g, a cherry red–oxford blue transformation is easily noticeable for polythiophenes.

C. Visy, *In situ Combined Electrochemical Techniques for Conducting Polymers*, SpringerBriefs in Applied Sciences and Technology, DOI 10.1007/978-3-319-53515-9_2

2.3   Mass Change

During the electrochemical transformations, the charge of the polymer backbone is changing. During the oxidation, the electron withdrawal creates excess positive charges. As a consequence of the electroneutrality rule, charge compensation through ion exchange with the adjacent solution should occur. This process is called improperly *doping* for virtual similarities in the conductance increase of inorganic semiconductors related to the presence of dopant, although there the cause and the consequence are just opposite.

Theoretically, there are two possibilities for the charge compensation process, due primarily to the relative ion mobilities within the channels of the given film. For the oxidative transformation of the polymer, the most evident consequence is the uptake of anions, which causes a mass increase. In other cases the electrochemically generated positive charges along the polymer chains expel cations from the film, and its mass is decreasing. Evidently, this may happen only when cations were already present in the CP in its neutral form, i.e., this layer contains a certain amount of both positive and negative ions with equal total charges. In several cases, when the mobility of the different ions is comparable, the motion of both ions can be revealed, and two distinct or overlapping sections in the mass change can be distinguished. During the discharging process the movement of the ions is opposite, and during a reversible redox cycle, the original mass of the film is regained. However, cycling the film in a solution containing other ions than those which were present during the electropolymerization process, the ion content of the layer can be modified. In that sense, we can distinguish CPs with anion or cation exchanger properties.

In special cases when some part of the polymer is able to dissociate, the so-called self-doping occurs via deprotonation (e.g., carboxyl derivatives). The removal of protons may take place also during the vigorous oxidation (from emeraldine base to pernigraniline) of polyaniline. The self-doping connected to proton movements causes small mass changes, generally close to the sensitivity limit of the measurement.

2.4   Conductance Change

A phenomenologic effect of the conductivity change manifests in polymer-based transistors which are already on the market. An organic field-effect transistor (OFET) uses an organic semiconductor in its channel. The most commonly used device geometry is bottom gate with top drain and source electrodes similar to the thin-film silicon transistor. The source and drain electrodes are directly deposited onto the thin layer of semiconductor, then a thin film of insulator is deposited between the semiconductor and the metal gate contact. If there is zero bias, there is no carrier movement between the source and drain. When a proper oxidation potential is applied, a highly conductive channel forms at the interface.

Such a configuration can be used as an organic light-emitting device (OLET). The device structure comprises interdigitated gold source and drain electrodes, Positive (holes) as well as negative charges (electrons) are injected from the gold contacts into this layer leading to electroluminescence.

2.5 Diamagnetic/Paramagnetic Changes

Since the first product of the electrochemical oxidation, when one electron from the aromatic ring system has been withdrawn, is the monocation/polaron which is a cation radical, its paramagnetic pattern can be detected by ESR technique. When dications or bipolarons are formed, these species—possessing no unpaired electron —are not ESR active any more. Thus modification in the ESR signal during the electrochemical perturbation offers excellent tracing of the amount of intermediates. Although phenomenological consequence of the appearance/disappearance of paramagnetic species cannot be revealed, a special instrumental in situ technique— as it will be discussed in the Sect. 3.1—is a valuable tool for the discovery of the details.

2.6 Volume Changes

Since "doping" ions move together with their solvation sphere, ion incorporation may lead to the accumulation/depletion of the solvent in the channels, leading to swelling or shrinking of the surface layer. The process depends not only on the solvation properties of the ions, but considerably also from the wetting character- istics of the polymer in the actual solution. This phenomenon can result in bending of bilayer films of different solvophylic character, which motivated researchers to create artificial muscles.

2.7 Structural Changes

Incorporation or removal of ions (charged) and molecules (neutral) components has an effect other than simple swelling or shrinking of the layer: during a cyclic voltammetric process there is a continuously changing charge distribution within the film, which alters its polarizability, and—if it takes place in a modulated electric field—it manifests in structural changes connected also to the change in the dielectric/capacitive properties. Structural changes are connected to alteration in the extension of coplanarity along the polymeric backbone.

2.8   Transport Processes in the Adjacent Solution Layer

Income or outcome of the charge compensating ions results in concentration changes in the solution phase adjacent to the film, which in turns manifest in modification in the refractive index of this thin solution layer. Although this effect invisible to the naked eye, it can be visualized by special optical instrumentation (see Sect. 3.1).

The complexity of the phenomena related to the electroactivity of conducting polymers establishes their potential and already even realized applications. The importance of the field can be well illustrated by the fact that the 2000 year Nobel Prize in Chemistry has been awarded to Hideki Shirakawa, Alan MacDiarmid and Alan Heeger for the discovery and development of conductive polymers.

A special issue of the Synthetic Metals has been dedicated to the Nobel Laurates on the occasion of awarding [1].

Comprehensive overviews of these materials and their properties have been given in the multiple editions of the *Handbook of conducting polymers* [2–4], as well as in a later monograph [5].

References

1. Mele G, Epstein A (2001) Foreword. Synth met 125:1-1. doi:10.1016/S0379-6779(01)00506-9
2. Skotheim TA (ed) (1986) Handbook of conducting polymers. Dekker, New York
3. Skotheim TA, Elsenbaumer RL, Reynolds (eds) (1997) Handbook of conducting polymers. Dekker, New York-Basel
4. Skotheim TA, Reynolds (eds) (2007) Handbook of conducting polymers. CRC Press, London-New york
5. Inzelt G (2008, 2012) Conducting polymers. In: Scholz F (ed) A new era in electrochemistry, Springer, Berlin

Chapter 3
Overview of in situ Combined Electrochemical Techniques

The aim of each in situ electrochemical method is to monitor one of the above-presented consequences of the electrochemical process. The additional signal can be correlated with the perturbation on the same time scale, uncovering this way the relation between the electrochemical and the triggered secondary events.

3.1 Single in situ Electrochemical Methods

3.1.1 Coupled Color Changes—Spectroelectrochemistry in the UV-VIS-NIR Region

The pioneers of the application of the in situ UV-Vis spectroelectrochemistry established the possible mechanisms of the electrochemical deposition. They also underlined how to distinguish between the formation of a faintly and a properly oxidized film, which latter is indispensable for a continuous deposition. They also revealed the formation of various, optically differentiated charge carriers during the redox transformation, leading to the polaron—bipolaron/quasi-metallic state mechanism [1–5].

Soon after these fundamental experiments also the theoretical interpretation—based on the band theory of semiconductors has been formulated [6].

Thus, importance of the in situ spectroelectrochemical investigations can be evidenced during both the polymerization and the redox transformations in a monomer-free electrolyte. In each case, the working electrode is transparent and conducting—generally an indium tin-oxide (ITO) coated quartz glass plate. In the most frequent configuration, ITO electrode constitutes one of the walls of the cell.

Instead of a spectrophotometer equipped with monochromator, a diode array instrument able to acquire spectra repeatedly with a resolution of 2–3 nm in the UV-Vis-NIR wavelength region is applied. During the electropolymerization,

© The Author(s) 2017   7
C. Visy, *In situ Combined Electrochemical Techniques for Conducting Polymers*,
SpringerBriefs in Applied Sciences and Technology,
DOI 10.1007/978-3-319-53515-9_3

spectra are repetitively taken, and an increase in the absorbance in the whole visible range indicates the propagation of the deposition.

An array at characteristic wavelengths may represent how the deposition advances versus either the time elapsed or the charge consumed. Eventual appearance of an induction period instead of strict linearity from the origo reflects delayed deposition due to the formation of soluble oligomers.

For the redox transformation of an electroactive film, the most informative in situ spectroelectrochemical method is "voltabsorptiometry," when spectral measurements are combined with cyclic voltammetry [7].

From the spectra absorbance, arrays were taken at characteristic wavelength values, resulting in absorbance–potential data pairs. Their derivatives represent the deconvoluted absorbance spectra, and these voltabsorptiometric curves give the change in the rate of the formation or consumption of the given species with a shape, resembling a voltammogram.

New methods treating the spectral results were also developed: the differential evolutionary factor analysis (DEFA), which is sensitive to changes in the composition of the mixture during the evolution process, and the projection matrix method, which can be used to annihilate the influence of a known spectrum from a set of measured spectra in order to obtain information about the unknown components [8].

Spectroelectrochemical studies combined with matrix rank analysis were used to determine the number of independent species necessary for the description of the spectral observations [9].

A brief overview of spectroelectrochemical methods was presented with the purpose to be a guide for the novice in this field searching for one or several methods suitable for tackling a given problem. Various examples including several aspects of intrinsically conducting polymers were summarized in order to illustrate the great potential of a combined use of carefully selected spectroelectrochemical methods [10].

A summary of the available in situ spectroelectrochemical methods, their basic principles, their typical applications, and their limitations have also been reviewed [11].

3.1.2 Changes in the Chemical Bonds' Nature—in situ FTIR and RAMAN Spectroscopy

Transmission infrared peaks of polyacetylene have been assigned in Shirakawa and Ikeda's laboratory already in 1978 [12].

Since the isomerization of the *cis* linkage to the *trans* occurs during the Raman measurements, the Raman spectrum of "purely cis" polyacetylene could not be obtained. The Raman $v_{C=C}$ frequency of the *cis* sequences was almost independent on the excitation wavelength in contrast to what was observed for the $v_{C=C}$ frequency of the *trans* sequences.

Soon after, Fourier transform infrared photoacoustic (FTIR-PA) spectroscopy has been used to obtain the first published infrared spectrum of a heavily n-doped polyacetylene, and comparisons were made with the FTIR-PA spectrum of undoped polyacetylene infrared peaks previously assigned by Shirakawa and Ikeda. Moreover, spectral shifts and intensity differences between the n-doped and undoped polyacetylene were set against changes noted previously in the infrared spectra of p-doped polyacetylenes by MacDiarmid and Heeger. [13].

A novel IR spectroelectrochemical cell which allows for electrochemical synthesis and electrochemical analyses of thin pristine polypyrrole films, while rigorously protecting these films from air have been designed by Lei, and FTIR spectra—registered during the reduction of polypyrrole in transmission mode—have been followed to show the alteration of the N–H bond stretching [14].

However, the most frequent realization of this combined technique is the detection of vibrational modifications in the reflection mode. Fourier transform infrared spectroscopy study of an electrochemically synthesized polypyrrole film was conducted by means of reflection absorption spectroscopy (FTIR RAS) and the FTIR KBr disk transmission method. The observed IR spectrum profile for the ring-stretching band and the CH in-plane and out-of-plane modes for polypyrrole were found different from that of the monomer, reflecting the α-α' conjugated polymer structure [15].

The experimental setup of two different methods—external reflection absorption spectroscopy and attenuated total reflection spectroscopy—was also described and applied for polyaniline [16].

As a summary, we may state that generally in the case of CPs with non-degenerate ground state, the aromatic—quinoidal shifts during the oxidation/reduction processes can be easily detected in the IR region.

3.1.3 Diamagnetic/Paramagnetic Changes—in situ ESR Spectroscopy

During the application of the simultaneous electrochemical electron spin resonance—SEESR—technique, the electrochemical cell has to be placed in the hole of an apparatus generating magnetic field. An improved coaxial cell for in situ electrochemical ESR has been presented [17].

Owing to the disruption or rearrangement of the conjugating aromatic electron system during the redox processes, appearance or disappearance of paramagnetic species has been found of prime importance in uncovering the mechanism of the process. Already in the early period of the studies, the method was able to show that in polythiophenes the appearance of paramagnetic species is only temporary, while diamagnetic forms of the oxidized polymer are more stable, and the "polythiophene puzzle" has been started to be set in by evidencing two oxidation levels [18].

3.1.4 In situ NMR Spectroscopy

This in situ technique has been applied by Genies for the first time [19].

The differences in the EPR spectra were interpreted using the theory concerning the Curie and Pauli spins, and the influence of ions on the nature of the spins were interpreted in the light of ionic interactions between the positive charges of the solvated polymer chain and the solvated anions of the solution.

Soon after, in situ NMR spectroscopy became a useful method to study soluble conducting polymers: Poly(*omicron*-propylaniline), poly(*omicron*-hexylaniline) proved to be more soluble than polyaniline. The higher solubility made possible the NMR study in solution, moreover it conferred a better processability on this new polymer [20].

Although spectroscopy was widely applied to study proton conducting polymer membrane, ex situ ^1H and ^{13}C NMR spectroscopy has to be referred here for its clinching for the synthesis of structurally homogenous, regioregularly well-defined alkylthiophene polymers [21].

3.1.5 Coupling with Mass Change—the Electrochemical Quartz Crystal Micro/Nanogravimetry (EQCM or EQCN) and the Radiotracer Technique

The electrochemical quartz crystal nanobalance is an electrode which reflects the modification of its mass by a frequency change. According to Sauerbrey's equation, the mass change per unit area of the electrode is related to the frequency change, and the proportionality can be calculated, and confirmed by calibration. The base frequency of the (metal covered) crystal has to be large, generally 5 or 10 MHz. The technique is useful not only in determining the electrochemical equivalent of deposited films, but its sensitivity makes possible to measure the mass changes, due to charge compensation coupled with the ionic motions during the redox processes.

EQCM technique for other surface layers than conducting polymers has been applied already from the 1980s [22–25].

The converse piezoelectric effect, in which an electric field applied across a piezoelectric material induces a stress in that material, has spurred many recent developments in mass measurement techniques. Advances in the methodology allowed dynamic measurements of mass changes $<10^{-9}$ grams per square centimeter at surfaces, thin films and electrode interfaces in liquid media as well [26].

Later the data handling became more sophisticated, making possible to distinguish and separate the contributions of different species, i.e., ion and solvent exchanges between the film and the adjacent solution [27].

The motions of these two species were in opposite directions, resulting in nearly compensatory mass changes. Comparison of the mass and charge fluxes showed that the two were correlated during PBT oxidation, but not during reduction, as the reduction of doped polymer occurred in two stages.

Although the original formula of Sauerbrey was formulated for the adsorption of gases, its validity was confirmed for the heterojunction of an electrode provided the surface layer can be considered rigid, not viscoelastic. These effects have been revealed and discussed a bit later compared to the first applications of the method, thus leading to the necessity of the reconsideration of several former conclusions [28].

In their work, they explored a new general transmission line model that described the viscoelastic characteristics of thin films in terms of their shear moduli. For the electrodeposition of poly(2,2'-bithiophene) conducting polymer films they came to the conclusion that the polymer behaved as a rigidly coupled mass only in the very early stages of deposition, beyond this it was a viscoelastic material.

An overview of the method has been compiled by Inzelt [29].

The electrochemical nanogravimetry is frequently used to determine the average molar mass of the in situ formed (doped) monomer unit, where the determination of the slope of the f frequency—Q charge curve (or vs. time curve in the galvanostatic mode) as well as an independent experimental information or guess for the oxidation level of the film are needed.

For the calculation of the apparent molar mass (M) of the depositing species the following equation can be used:

$$M = nFA\delta f / C_f Q,$$

where n (= 2 + d) is the number of electrons transferred in the deposition reaction, d is the doping level, F is Faraday constant, A is the acoustically active surface area, δf is the frequency change, C_f is the constant for the crystal in the Sauerbrey equation, and Q is the charge consumed.

During the application of **the radiotracer method**, the electrode in form of a foil, constituting the bottom of the cell, is placed on the window of the radioactivity detector. Using labeled anions in the electrolyte, their accumulation or depletion in the polymer (in the close vicinity of the counter of β radiation) could be adequately distinguished from the (constant) background radioactivity [30].

It is a plausible consequence of the method that the information is not falsified or distorted by solvation, which has to be encountered when using EQCM [31, 32].

3.1.6 Resulted Conductance Change—a.c. Conductance, Electrochemical Impedance Measurements

This in situ electrochemical technique was first applied in Murray's laboratory, and developed further by Musiani [33, 34].

In order to measure the conductance of the organic layer under electrochemical control, a double band working electrode is needed. However, the insulating gap between the bands must not exceed 10–20 μm, a distance which is small enough to insure the two parts of growing polymer to be interconnected at a later stage of the film growth. The two, originally separated supporting electrodes serve then also as contacts for the conductance measurements.

The similar arrangement can be realized on a printed circuit electrode, some types of which are commercially available.

On understanding that the gap width and the thickness of the deposited layer are commensurable, the conductivity σ can be determined from the (linear) slope of conductance G versus logarithm of charge Q curves according to the equation

$$G = A + (\sigma * 1 / \pi) * \ln Q,$$

where l is the length of the double band electrodes [35].

In situ a.c. conductance measurements means that an alternating sine wave of properly chosen amplitude and frequency is superimposed on the potential profile. The resulted current response is encumbered by this perturbation. The measured impedance is then decomposed either to its real and imaginary components or to its absolute value and phase shift using a lock-in amplifier.

Performing the measurements in a wide frequency range, the electrochemical impedance spectrum can be registered, which is used to give the equivalent circuit adequate with the electric properties of the system, and to deduce the value of the parameters characterizing the system such as solution resistance, double layer capacity, charge transfer resistance, Warburg impedance, etc.

A comprehensive overview has been given by G. Inzelt, and G. G. Láng [36].

3.1.7 Effects Monitored in the Solution—Laser Beam Deflection

This technique is based on the fact that ionic motions, due to charge compensation, cause inhomogenity at the polymer–solution interface in the vicinity of the polymer. The resulted concentration gradient alters the refractive index of the solution layer. A laser beam passing in this range is deteriorated, which can be detected using a quadruple photodiode, having positioned symmetrically beforehand.

The method was introduced in 1990 for studying oxide films [37].

The phenomenon, called also probe beam deflection or "mirage effect," can be characterized by the deflection angle, changing in parallel with the flux, being opposite in sign when originating from cation and anion movements. Properly applied mathematical treatment may lead to a mirage deflection curve with similarities to the actual voltammogram [38].

3.1.8 Subsequent Volume Change

When followed by AFM, the technique is able to reveal the redox transformation coupled swelling or shrinking of the film, due assumingly to the movement of solvated ions, and structural modifications in a changing electric field [39].

The method was used also to characterize the roughness by the fractal dimension analysis and to make comparisons of the morphology changes of polypyrrole films deposited onto various (polycrystalline platinum, glassy carbon, and gold) electrodes during both the synthesis and the oxidation–reduction processes.

The effect of the electrochemical transformation on the volume of the film can be studied also by in situ reflection spectroscopy [40].

3.1.9 Visualization of Structural Changes: Ellipsometry, Surface Plasmon Resonance, X-Ray Absorption Fine Structure Spectroscopy (EXAFS)

The ellipsometric technique for studying electroactive polymers has been introduced by Hillman et al. [41].

The combination made possible the quantitative correlation of electrochemical and ellipsometric data for thin films, in situ monitoring of the growth of thin electrochemically polymerised films on rapid timescales as well as the direct determination of the solvent content of the film.

The principle of plasmonic imaging of electrochemical currents using electrochemical-surface plasmon resonance for a model system was described [42].

The quantitative formalism of electrochemical-surface plasmon resonance (EC-SPR) provided a new way to measure convolution voltammetry without the need of numerical integration.

In a recent work, implementation of electrochemistry with plasmonic nanostructures for combined electrical and optical signal transduction has been achieved. As it was demonstrated, the redox reaction caused a local change in refractive index because of the reactants consumed and products generated. By imaging surface plasmon resonance, one could visualize where the reaction actually occurred [43].

As for the use of extended X-ray absorption fine structure spectroscopy to probe electrode–solution interfaces, a review of the development and application of EXAFS for in situ structural characterization of chemical processes at the electrode surface was given, and the different designs of the developed in situ electrochemical cells were presented. Moreover, applications were selected from the areas of metal passivation, spectroelectrochemistry, modified electrodes, underpotential deposition, and adsorption [44].

3.2   Combination of Two Single in situ Electrochemical Techniques in a Hyphenated Mode

In the previous chapter, the complementary pattern of in situ techniques, applied under electrochemical circumstances was evidenced. Numerous experts working in the field of CPs make efforts to collect complex information about the behavior of the given film in order to get an even more complete picture from the actual processes. Thus, more and more examples are available for the combinations of two in situ techniques in a hyphenated mode, where multifaceted characterization of the self-same layer can be achieved.

It is evident that massing of these methods is limited by either the technical realization or the rationality. Measuring, let us say, by the atomic force microscopy can be hardly harmonized with laser beam deflection, since the first makes the second impossible. Till the moment, until these restrictions should be taken into consideration, conclusions from measurements performed on identically prepared polymer samples can only be drawn. As it has been proved in the previous chapters, even parallel treatment of the results obtained by different but singularly used in situ methods can step forward in the deeper and deeper elucidation of the details.

The increment of hyphenating two in situ combined electrochemical techniques is to gain information on the same phenomenon for the self-same polymer film from multiple aspects at the same time. Driven by the electrochemistry, the two in situ methods deliver data independently from each other, but both data sets are the direct consequence of the identical electrochemical perturbation. This common background makes possible to correlate the arrays of independent signals, and to reveal the togetherness or the separation of the different yields of the electrochemical effect.

3.2.1   Simultaneous Combination of the Electrochemical Electron Spin Resonance (SEESR) and the in situ a.c. Conductance Techniques

Modern research in spectroelectrochemistry is characterized by a combination of different spectroscopic methods for a detailed study of electrode reaction mechanisms or complex electrode systems. Considering the reaction of a neutral organic redox system, it was stated already in the introduction chapter that the primary cathodic or anodic electron transfer would result in an ion radical. Applying a double-band electrode in a cell, which can be inserted into the hole of the ESR magnetic field, the combination may correlate the development of the macroscopic conductance with the generation of paramagnetic species [45].

When the electrochemical doping of polypyrrole was studied, a quantitative estimate of the polypyrrole conductivity as a function of doping level has been obtained and the corresponding change in the polaron concentration has been measured. It was found that at low doping levels there was a good correlation between these two properties. It was suggested that the polaron may be responsible for the conduction process under such conditions. At higher doping levels conduction appeared to involve the bipolaron.

3.2.2 Simultaneously Applied Electron Spin Resonance (SEESR) and in situ Spectroelectrochemistry

The ESR spectroscopy, enabling to prove the existence of a paramagnetic structure, can be combined by other spectroscopic methods. The technical solution of this combination is achieved using a transparent electrode in the cell, which has to be introduced into the magnetic field. When applying these two in situ methods in a synchronized manner, paramagnetic and diamagnetic species can be identified in the UV-Vis-NIR spectra. This hyphenation may discover the pattern of the interaction of radical species generated in the electrochemical transformation, as it was described by Dunsch et al. [46].

Here, simultaneous in situ measurements by both ESR and UV-vis spectroscopy have been carried out at the same electrode for the first time during a single electrochemical experiment in order to spectroscopically characterize the electrochemical reaction products. The experimental technique, including a special optical ESR cavity and an electrochemical cell for both ESR and UV-vis spectroscopy in transmission, was described. It was shown that the UV-vis absorbance measured in situ, as well as the ESR intensity characteristics of the electrochemical system under study, can be understood in terms of the faradaic current. The reliability of this system has proved by measuring the organic redox couple of methyl-substituted p-phenylenediamine and by comparing calculated and experimental curves.

3.2.3 Combination of in situ UV-Vis-NIR Spectroelectrochemical and in situ a.c. Conductance Measurements

For this hyphenation a transparent double-band electrode is needed. The organic layer is deposited onto the pair of electrodes consisting of ITO. In order to control both the conductance and the spectral measurements, two computers have to be synchronized [47].

As it was proved, this combination of the two in situ electrochemical techniques was able to deliver direct evidence for what an extent a given, spectrally distinguishable species is responsible for the evolution of the *name-giving* property of conducting polymers.

3.2.4 In situ Piezoelectric Spectroelectrochemisty

A. Electrogravimetry combined with in situ UV-vis *reflectance mode*

The first application of the combination of these two in situ techniques was presented by Shimazu et al., although not for a conducting polymer film [48].

The sensor crystal used was an AT-cut quartz plate with patterned ITO, insuring this way the acquisition of both mass and spectral datasets.

The use of the method augurs well for widespread applications not only in the field of conducting polymers, but also in any other electrochemical experiments where colored (or light emitting) species appear [49].

Here a software-controlled electrochemical measuring system has been developed, and its capabilities for studying electropolymerization reactions were demonstrated. Based on the corresponding adsorption and frequency changes yielded by the measured multidimensional dataset, the analysis of a special extinction coefficient (molar absorptivity) was presented together with a promising application.

B. Electrogravimetry combined with fiberoptic reflectance

An instrumental setup has been configured for a simultaneous real-time recording of electrochemical, spectroelectrochemical, and mass data on an electrode from a single measurement for studies of complex electrochemical reaction mechanisms. This was achieved by combining a potentiostat/galvanostat for electrochemical measurements, the near-normal incidence reflectance spectrometer for spectroelectrochemical measurements, and a quartz crystal analyzer (QCA) for mass changes on the electrode. The latter two were coupled through a bifurcated optical fiber on the reflective QCA electrode [50].

C. Electrogravimetry combined with FTIR

A hyphenation of the electrochemical quartz crystal nanogravimetry and reflectance mode IR spectroscopy was achieved. Using this hyphenation, electrodeposition of polymer films could be followed by the identification of possible chemical bonds. When the metal-coated EQCM electrode is covered by the polymer, development of various bonds could be monitored, and alterations could be related to the amount of the deposited material [51].

D. Electrogravimetry combined with ellipsometry

Combined measurements performed during the anodic growth of polyaniline in aqueous acid solutions have been described. The information obtained from the two techniques is complementary: the QCM measured the overall mass increase during film growth, whereas the ellipsometer measured the optical properties of the film and its thickness. From such combined simultaneous measurements, the apparent density of the film could be derived [52].

E. Electrogravimetry combined with Scanning Electrochemical Microscopy

The combined scanning electrochemical microscope-quartz crystal microbalance instrument for studying thin films has been presented by Cliffel and Bard [53].

The SECM−QCM was used to study etching of a thin Ag layer deposited on the QC contact by generating an etchant, iron(III) tris(bipyridine), at the tip near the surface. The SECM–QCM was also used to monitor film mass and surrounding electrolyte composition during potential cycling of a film of C60 on an electrode.

The capability of the EQCM/SECM instrument was tested in cyclic voltammetric plating/stripping experiments. The amperometric response of the SECM tip positioned closely to the substrate reflected the concentration changes of electroactive ions in the diffusion layer [54].

The combination of scanning electrochemical microscopy (SECM) with piezoelectric quartz crystal impedance (PQCI) analysis was proposed as a novel multiparameter method for investigating the cyclic voltammetric growth of poly (o-phenylenediamine) thin films at electrodes in aqueous solutions, and for comparing the polymer porosity and stability during the potentiostatic microetching of these films [55].

F. Electrogravimetry combined with Probe Beam Deflection PBD

This hyphenation is useful to distinguish and separate ionic and solvent movements, applied for studying the electrodeposition of polyaniline and its derivatives [56].

During the first oxidation step, the positive sign of the beam deflection clearly indicated ion insertion into both films from the electrolyte. Ion insertion was indicated also by a frequency decrease (mass increase). The quantitative evaluation of the slope of the frequency-charge plots lead to an effective molar mass comparable well with the molar mass of the anion.

A frequency increase (mass decrease) found for the second step, indicated ion expulsion involving also proton removal. The blocking of the imine nitrogen in polyanilines by methyl groups affected the second oxidation step, since deprotonation of the imino groups was no longer possible, and consequently only anions were exchanged during the second oxidation step.

Although the model was metal deposition, the electrochemical quartz crystal microbalance (EQCM) and probe beam deflection (PBD) have been used to study interfacial processes in room temperature ionic liquids, including deep eutectic solvents (DES) [57].

While overcoming the effect of viscous loss in the acoustic wave (EQCM) part of the experiment could be understood, the optical (PBD) technique failed to provide a meaningful response in slow scan rate voltammetric experiments, in contrast with those obtained in aqueous media. As it was proved, the problem could be overcome by operating at scan rates 1–2 orders of magnitude slower. The solution of the problem permits the application of this powerful technique to novel media of technological interest.

G. **Electrogravimetry combined with Electrochemical Impedance Spectroscopy (EIS)**

Although this hyphenation has been applied to study the underpotential deposition of various noble metals, its potential in the electrodeposition process of CPs is self-evident [58].

H. **Electrogravimetry combined with Surface Plasmon Resonance**

The advantage of the combination of surface plasmon resonance and the quartz crystal microbalance is widescale. This hyphenation has been applied to test the validity of the governing equations often used to analyze data collected using the two techniques, bringing to light weaknesses in the assumptions inherent in these equations. The results were used to calculate interfacial density and refractive index values in a system where the bulk values are known and the physical state of the adsorbed material is similar to that of the bulk, to show that the viscoelastic properties of an adsorbed material change significantly as the material desorbs from an interface, and to observe the evolution in the electronic and chemical properties of a conducting polymer film as it is being deposited while precisely monitoring the mass of the deposited film [59].

Electropolymerization and doping/dedoping properties of polyaniline thin films was studied by electrochemical-surface plasmon spectroscopy and by the quartz crystal microbalance [60].

The combination of electrochemical surface plasmon resonance (ESPR) spectroscopy and the electrochemical quartz crystal microbalance have been applied at two wavelengths and as a function of the applied potential. The real and imaginary parts of the dielectric constant of the polyaniline at several doping levels was determined quantitatively by taking into consideration the thickness values obtained from the EQCM measurement.

3.2.5   *In situ FTIR Spectroelectrochemistry and Ellipsometry*

Results from in situ ellipsometry and Fourier transform infrared spectroscopy on the growth, the electrochemical cycling and the overoxidation of polypyrrole showed that the initial charge-carrying species are polarons. At higher potentials, bipolarons were also formed, and coexisted with the polarons strongly suggesting that there is no appreciable energy gain in forming a bipolaron (with respect to two polarons). Both species have narrow, well-defined conjugation lengths, with bipolarons being ca 9 monomer units smaller than polarons. The oxidation of the polypyrrole results in the expulsion of solvated protons from the film. Importantly, oxidation of the film is accompanied by a reversible 30% reduction in its thickness, a quite unexpected result in a view of the necessity for charge compensation through ingress of solvated anions [61].

3.2.6 Combined Scanning Electrochemical—Atomic Force Microscopy AFM

The schematic setup for this combination was described in 2000. The instrument made possible making the first simultaneous topographical and electrochemical measurements at surfaces under fluid with high spatial resolution. From technical point, the simple probe tips suitable for SECM-AFM have been fabricated by coating flattened and etched Pt microwires with insulating, electrophoretically deposited paint [62].

3.2.7 Current-Sensing Atomic Force Microscopic (CS-AFM) and Spectroscopic Measurements

Layout of the current-sensing AFM and the near-normal incidence reflectance spectroscopy (NNIRS) setup for obtaining absorption spectra from the conducting polymer film has been developed for poly(3-methylthiophene) films [63].

By applying this hyphenation, the initial stages of the electropolymerization of poly(3,4-ethylenedioxythiophene) films on gold electrode were also studied simultaneously. In addition to the electrochemical information obtained, the surface plasmon optical technique showed the deposition, and provided electrochromic behavior of the film. At the same time, atomic force microscopy gave surface morphological properties and domain growth of the film. The complementarity of the information obtained from the combined techniques is useful for the investigation of polymer growth mechanism [64].

3.2.8 Combining Scanning Electrochemical Microscopy with Infrared Attenuated Total Reflection Spectroscopy

The capabilities of the SECM-IR-ATR were demonstrated by simultaneous spectroscopic monitoring of the microstructured electropolymerization which was induced by feedback mode SECM. The level of polymerization could be evaluated by the recorded IR spectra, which were synchronized with the progression of the electrochemically induced polymerization. Furthermore, access to the fingerprint region of the IR spectrum (>10 μm) not only revealed the polymerization level but provided direct spectroscopic insight on the polymerization mechanism [65].

3.2.9   Combined Electrochemical Impedance EIS and FTIR–ATR

This combined approach allowed the analysis of diffusion coefficients of water with two independent measurement techniques, the quantification of the water uptake and its influence on the interphasial composition. This combination has been used to monitor also the solvent movement within a polymer film [66].

References

1. Garnier F, Tourillon T, Gazard M, Dubois JC (1983) Organic conducting polymers derived from substituted thiophenes as electrochromic material. J Electroanal Chem 148:299–303. doi:10.1016/S0022-0728(83)80406-9
2. Genies EM, Bidan G, Diaz AF (1983) Spectroelectrochemical study of polypyrrole films. J Electroanal Chem 149:101–113. doi:10.1016/S0022-0728(83)80561-0
3. Kobayashi T, Yoneyama H, Tamura H (1984) Polyaniline film-coated electrodes as electrochromic display devices. J Electroanal Chem 161:419–423. doi:10.1016/S0022-0728 (84)80201-6
4. Kuwabata S, Yoneyama H, Tamura H (1984) Redox behavior and electrochromic properties of polypyrrole films in aqueous-solutions. Bull Chem Soc Jpn 57:2247–2253. doi:10.1246/ bcsj.57.2247
5. Inganas O, Lundstrom I (1984) A photoelectrochromic memory and display device based on conducting polymers. J Electrochem Soc 131:1129–1132. http://apps.webofknowledge.com/ full_record.do?product=WOS&search_mode=GeneralSearch&qid=1&SID=Q1qEYwxHJ133 rClpT9s&page=1&doc=1. Accessed 5 Dec 2016
6. Brédas JL, Themans B, Fripiat JG, Andre JM, Chance RR (1984) Highly conducting polyparaphenylene, polypyrrole and polythiophene chains—an abinitio study of the geometry and electronic-structure modifications upon doping. Phys Rev B 29:6761–6773. doi:10.1103/ PhysRevB.29.6761
7. Lukkari J, Kankare J, Visy C (1992) Cyclic spectrovoltammetry: a new method to study the redox processes in conductive polymers. Synth Met 48:181–192. doi:10.1016/0379-6779(92) 90060-V
8. Kankare J, Lukkari J, Pajunen T, Ahonen J, Visy C (1990) Evolutionary spectral factor analysis of doping-undoping processes of thin conductive polymer films. J Electroanal Chem 294:59–72. doi:10.1016/0022-0728(90)87135-7
9. Visy C, Krivan E, Peintler G (1999) MRA combined spectroelectrochemical studies on the redox stability of PPy/DS films. J Electroanal Chem 462:1–11. doi:10.1016/S0022-0728(98) 00384-2
10. Holze R (1994) Spectroscopic methods in electrochemistry - new tools for old problems. Bull Electrochem 10:45–55. http://apps.webofknowledge.com/full_record.do?product=WOS& search_mode=GeneralSearch&qid=6&SID=Q1qEYwxHJ133rClpT9s&page=1&doc=2. Accessed 5 December 2016
11. Plieth W, Wilson GS, De La Fe CG (1998) Spectroelectrochemistry: a survey of in situ spectroscopic techniques. Pure Appl Chem 70:1395–1414. doi:10.1351/pac199870071395
12. Harada I, Tasumi M, Shirakawa H, Ikeda S (1978) Raman-spectra of polyacetylene and highly conducting ione-doped polyacetylene. Chem Lett 12:1411–1414. doi:10.1246/cl.1978.1411
13. Riseman SM, Yaniger SI, Eyring EM, Macinnes D, MacDiarmid AG, Heeger AJ (1981) Infrared photo-acoustic spectroscopy of conducting polymers 1. Appl Spectrosc 35:557–559. doi:10.1366/0003702814732085

14. Lei JT, Liang WB, Martin CR (1992) Infrared investigation of pristine, doped and partially doped polypyrrole. Synth Met 48:301–312. doi:10.1016/0379-6779(92)90233-9

15. Kato H, Nishikawa O, Matsui T, Honma S, Kokado H (1991) Fourier transform infrared spectroscopy study of conducting polymer polypyrrole: Higher order structure of electrochemically synthesized film. J Phys Chem 95:6014–6016. doi:10.1021/j100168a055

16. Neugebauer H (1995) In-situ vibrational spectroscopy of conducting polymer electrodes. Macromolecular symposia 94:61–73. doi:10.1002/masy.19950940107

17. Compton RG, Waller AM (1985) An improved cell for in-situ electrochemical ESR. J Electroanal Chem 195:289–297. doi:10.1016/0022-0728(85)80049-8

18. Zotti G, Schiavon G (1989) The polythiophene puzzle—electrochemical and spectroelectrochemical evidence for two oxidation levels. Synth Met 31:347–357. doi:10.1016/0379-6779 (89)90802-3

19. Lapkowski M, Genies EM (1990) Evidence of 2 kinds of spin in polyaniline from in situ EPR and electrochemistry. J Electroanal Chem 279:157–168. doi:10.1016/0022-0728(90)85173-3

20. Genies EM, Noel P (1991) Synthesis and polymerization of ortho-hexylaniline—characterization of the corresponding polyaniline. J Electroanal Chem 310:89–111. doi:10.1016/0022-0728(91)85254-M

21. McCullough RD, Loewe RD, Jayaraman M, Anderson DL (1993) Design, synthesis, and control of conducting polymer architectures—structurally homogeneous poly(3-alkylthiophenes). J Org Chem 58:904–912. doi:10.1021/jo00056a024

22. Albery WJ, Foulds AW, Hall KJ, Hillman AR (1980) Thionine-coated electrode for photogalvanic cells. J Electrochem Soc 127:654–661. doi:10.1149/1.2129727

23. Nomura T, Okuhara M (1982) Frequency-shifts of piezoelectric quartz crystals immersed in organic liquids. Anal Chim Acta 142:281–284. doi:10.1016/S0003-2670(01)95290-0

24. Kaufman JH, Kanazawa KK, Street GB (1984) Gravimetric electrochemical voltage spectroscopy—in situ mass measurements during electrochemical doping of the conducting polymer Polypyrrole. Phys Rev Lett 53:2461–2464. doi:10.1103/PhysRevLett.53.2461

25. Bruckenstein S, Shay M (1985) Experimental aspects of use of the quartz crystal microbalance in solution. Electrochim Acta 30:1295–1300. doi:10.1016/0013-4686(85) 85005-2

26. Ward MD, Buttry DA (1990) In situ interfacial mass detection with piezoelectric transducers. Science 249:1000–1007. doi:10.1126/science.249.4972.1000

27. Hillman AR, Swann MJ, Bruckenstein S (1990) Ion and solvent transfer accompanying polybithiophene doping and undoping. J Electroanal Chem 291:147–162. doi:10.1016/0022-0728(90)87183-K

28. Bandey HL, Hillman Brown MJ, Martin SJ (1997) Viscoelastic characterization of electroactive polymer films at the electrode/solution interface. Faraday Discuss 107: 105–121. doi:10.1039/a704278g

29. Inzelt G (2010) Elecrochemical quartz crystal nanobalance. In: Scholz F (ed) Electroanalytical methods, Guide to experiments and applications, 2nd revised and extended edition. Springer, Berlin, vol II, no 10, pp 257–270

30. Inzelt G, Horányi G (1987) Combined electrochemical and radiotracer study of anion sorption from aqueous solutions into polypyrrole films. J Electroanal Chem 230:257–265. doi:10. 1016/0022-0728(87)80147-X

31. Horányi G, Inzelt G (1988) Application of radiotracer methods to the study of the formation and behavior of polymer film electrodes—investigation of the formation and overoxidation of labeled polyaniline films. J Electroanal Chem 257:311–317. doi:10.1016/0022-0728(88) 87052-9

32. Horányi G, Inzelt G (1988) Anion-involvement in electrochemical transformations of polyaniline—a radiotracer study. Electrochim Acta 33:947–952. doi:10.1016/0013-4686(88) 80093-8

33. Feldman BJ, Burgmayer P, Murray RW (1985) The potential dependence of electrical-conductivity and chemical charge storage of poly(pyrrole) films on electrodes. JACS 107:872–878. doi:10.1021/ja00290a024

34. Musiani MM (1990) Characterization of electroactive polymer layers by electrochemical impedance spectroscopy. Electrochim Acta 35:1665–1670. doi:10.1016/0013-4686(90) 80023-H
35. Kankare J, Kupila EL (1992) In situ conductance measurement during electro-polymerization. J Electroanal Chem 322:167–181. doi:10.1016/0022-0728(92)80074-E
36. Inzelt G, Láng GG (2010) Electrochemical impedance spectroscopy (EIS) for polymer characterization. In: Cosnier S, Karyagin A (eds) Electropolymerization. Wiley-VCH, Weinheim, Germany, pp 51–76
37. Kötz R, Barber C, Haas O (1990) Probe beam deflection investigation of the charge storage reaction in anodic iridium and tungsten-oxide films. J Electroanal Chem 296:37–49. doi:10.1016/0022-0728(90)87231-8
38. Vieil E, Matencio T, Plichon V, Servagent S (1991) Hysteresis, relaxation and ionic movements in conducting polymers studied by in-situ quartz microbalance, ESR and mirage effect. Synth Met 43:2837–2837. DOI:10.1016/0379-6779(91)91186-E
39. Silk T, Hong Q, Tamm J, Compton RG (1998) AFM studies of polypyrrole film surface morphology - I. The influence of film thickness and dopant nature. Synth Met 93:59–64. doi:10.1016/S0379-6779(98)80131-8
40. Hillman AR, Mallen EF (1991) Electroactive bilayers employing conducting polymers 2. Speciation by in situ spectroscopy. J Chem Soc, Faraday Trans 87:2209–2217. doi:10.1039/ft9918702209
41. Hillman AR, Hamnett A, Christensen PA (1987) Ellipsometry of polythiophene films on electrodes. J Electrochem Soc 134:C131–131. http://apps.webofknowledge.com/full_record. do?product=WOS&search_mode=GeneralSearch&qid=22&SID=Q1qEYwxHJ133rClpT9s& page=1&doc=1. Accessed 5 Dec 2016
42. Wang S, Huang X, Shan XN, Foley KJ, Tao NJ (2010) Electrochemical surface plasmon resonance: basic formalism and experimental validation. Anal Chem 82:935–941. doi:10.1021/ac902178f
43. Dahlin AB, Dielacher B, Rajendran P, Sugihara K, Sannomiya T, Zenobi-Wong M, Voros J (2012) Electrochemical plasmonic sensors. Anal Bioanal Chem 402:1773–1784. doi:10.1007/s00216-011-5404-6
44. Dewald HD (1991) Use of exafs to probe electrode solution interfaces. Electroanalysis 3: 145–155. doi:10.1002/elan.1140030303
45. Waller AM, Compton RG (1989) Simultaneous alternating-current impedance electron-spin resonance study of electrochemical doping in Polypyrrole. J Chem Soc Faraday Trans 1 F 85:977–990. DOI:10.1039/f19898500977
46. Petr A, Dunsch L, Neudeck A (1996) In situ UV-Vis ESR spectroelectrochemistry. J Electroanal Chem 412:153–158. doi:10.1016/0022-0728(96)04582-2
47. Peintler-Kriván E, Tóth PS, Visy C (2009) Combination of in situ UV-Vis-NIR spectro-electrochemical and a.c. impedance measurements: a new, effective technique for studying the redox transformation of conducting electroactive materials. Electrochem Commun 11:1947–1450. doi:10.1016/j.elecom.2009.08.025
48. Shimazu K, Yanagida M, Uosaki K (1993) Simultaneous UV-Vis spectroelectrochemical and quartz crystal microgravimetric measurements during the redox reaction of viologens. J Electroanal Chem 350:321–327. doi:10.1016/0022-0728(93)80214-3
49. Berkes BB, Vesztergom S, Inzelt G (2014) Combination of nanogravimetry and visible spectroscopy: a tool for the better understanding of electrochemical processes. J Electroanal Chem 719:41–46. doi:10.1016/j.jelechem.2014.01.031
50. Shim HS, Yeo IH, Park SM (2002) Simultaneous multimode experiments for studies of electrochemical reaction mechanisms: demonstration of concept. Anal Chem 74:3540–3546. doi:10.1021/ac020121c
51. Liu M, Ye M, Yang Q, Zhang Y, Xie Q, Yao S (2006) A new method for characterizing the growth and properties of polyaniline and poly(aniline-co-o-aminophenol) films with the combination of EQCM and in situ FTIR spectroelectrochemistry. Electrochim Acta 52: 342–352. doi:10.1016/j.electacta.2006.05.013

52. Rishpon J, Redondo A, Derouin C, Gottesfeld S (1990) Simultaneous ellipsometric and microgravimetric measurements during the electrochemical growth of polyaniline. J Electroanal Chem 294:73–85. doi:10.1016/0022-0728(90)87136-8

53. Cliffel DE, Bard AJ (1998) Scanning electrochemical microscopy. 36. A combined scanning electrochemical microscope quartz crystal microbalance instrument for studying thin films. Anal Chem 70:1993–1998. doi:10.1021/ac971217n

54. Gollas B, Bartlett PN, Denuault G (2000) An instrument for simultaneous EQCM impedance and SECM measurements. Anal Chem 72:349–356. doi:10.1021/ac990796o

55. Tu XM, Xie QJ, Xiang CH, Zhang YY, Yao SZ (2005) Scanning electrochemical microscopy in combination with piezoelectric quartz crystal impedance analysis for studying the growth and electrochemistry as well as microetching of poly(o-phenylenediamine) thin films. J Phys Chem B 109:4053–4063. doi:10.1021/jp044731n

56. Barbero C, Miras MC, Haas O, Kötz R (1991) Alteration of the ion-exchange mechanism of an electroactive polymer by manipulation of the active-site - probe beam deflection and quartz crystal microbalance study of poly(aniline) and poly(n-methylaniline). J Electroanal Chem 310:437–443. doi:10.1016/0022-0728(91)85280-3

57. Hillman AR, Ryder KS, Zaleski CJ, Ferreira V, Beasley CA, Vieil E (2014) Application of the combined electrochemical quartz crystal microbalance and probe beam deflection technique in deep eutectic solvents. Electrochim Acta 135:42–51. doi:10.1016/j.electacta.2014.04.062

58. Berkes BB, Maljusch A, Schuhmann W, Bondarenko AS (2011) Simultaneous acquisition of impedance and gravimetric data in a cyclic potential scan for the characterization of nonstationary electrode/electrolyte interfaces. J Phys Chem C 115:9122–9130. doi:10.1021/jp200755p

59. Bailey LE, Kambhampati D, Kanazawa KK, Knoll W, Frank CW (2002) Using surface plasmon resonance and the quartz crystal microbalance to monitor in situ the interfacial behavior of thin organic films. Langmuir 18:479–489. doi:10.1021/la0112716

60. Baba A, Tian S, Stefani F, Xia CJ, Wang ZH, Advincula RC, Johannsmann D, Knoll W (2004) Electropolymerization and doping/dedoping properties of polyaniline thin films as studied by electrochemical-surface plasmon spectroscopy and by the quartz crystal microbalance. J Electroanal Chem 562:95–103. doi:10.1016/j.jelechem.2003.08.012

61. Christensen PA, Hamnett A (1991) In situ spectroscopic investigations of the growth, electrochemical cycling and overoxidation of polypyrrole in aqueous-solution. Electrochim Acta 36:1263–1286. doi:10.1016/0013-4686(91)80005-S

62. Macpherson JV, Unwin PR (2000) Noncontact electrochemical imaging with combined scanning electrochemical atomic force microscopy. Anal Chem 72:276–285. doi:10.1021/ac990921w

63. Lee HJ, Park SM (2004) Electrochemistry of Conductive Polymers. 33. Electrical and optical properties of electrochemically deposited poly(3-methylthiophene) films employing current-sensing atomic force microscopy and reflectance spectroscopy. J Phys Chem B 108:16365–16371. doi:10.1021/jp0472764

64. Baba A, Knoll W, Advincula R (2006) Simultaneous in situ electrochemical, surface plasmon optical and atomic force microscopy measurements: Investigation of conjugated polymer electropolymerization. Rev Sci Instrum 77:064101. doi:10.1063/1.2204587

65. Wang LQ, Kowalik J, Mizaikoff B, Kranz C (2010) Combining scanning electrochemical microscopy with infrared attenuated total reflection spectroscopy for in situ studies of electrochemically induced processes. Anal Chem 82:3139–3145. doi:10.1021/ac9027802

66. Vlasak R, Klueppel I, Grundmeier G (2007) Combined EIS and FTIR–ATR study of water uptake and diffusion in polymer films on semiconducting electrodes. Electrochim Acta 52:8075–8080. doi:10.1016/j.electacta.2007.07.003

Chapter 4
Applications of the in situ Combined Electrochemical Techniques: Problems and Answers Attempted by the in situ Combined Methods

As the number of the contributions to the topic is quasi infinite compared to the length of this book, a restricted number of examples are selected in this chapter to demonstrate how in situ electrochemical methods can give the answer to diverse questions, arisen in connection with either the electropolymerization or the redox switching process.

4.1 Electrodeposition

The first steps toward the elucidation of the mechanism of the electrochemical deposition of conducting polymers have been made already in the 1980s by in situ UV-VIS-NIR spectroelectrochemical techniques [1–5].

On the basis of in situ infrared IR study, study, it could be concluded that polythiophene growth is accompanied by gross changes in the IR spectrum from that of the monomer in solution. Oxidation of the film gives rise to additional bands corresponding to the "quinonoid" form of the chain, and to a reduction in the frequency of the ring deformation modes compared to the neutral form. A broad electronic absorption band corresponding to bipolaronic excitation during the polythiophene growth was also seen [6].

During in situ electrochemical ESR investigations of the growth of one- and two-dimensional polypyrrole films, differences in the paramagnetic properties of the PPy films have been observed depending on the applied current densities. Furthermore, the spin concentration in the PPy film was not proportional to the injected charge. As a conclusion, two different mechanisms of film formation were assumed, depending on the applied current densities [7].

The original version of this chapter was revised: The typographical error has been corrected. The erratum to this chapter is available at 10.1007/978-3-319-53515-9_6

© The Author(s) 2017
C. Visy, *In situ Combined Electrochemical Techniques for Conducting Polymers*,
SpringerBriefs in Applied Sciences and Technology,
DOI 10.1007/978-3-319-53515-9_4

Thienyl silanes have been found as excellent precursors for the deposition of highly conducting polythiophenes [8].

Later, this work has been completed by extending the UV-Vis spectroelectro-chemical studies also to tetrathienyls of other carbon group elements (Si, Ge, Sn, Pb). The previously reported excellent conductance of tetrathienylsilane could be achieved only in nitrobenzene solution, the solvent used in Ref. [8], and the deter-mining effect of the nucleophilic character of the various solvents has been assumed. However, differences observed in solvents of about the same donocity could not be interpreted. As it was concluded, neutral radicals, formed during the homolysis of the different tetratienyls, may participate in hydrogen abstraction reaction in solvents possessing hydrogens bound to aliphatic chain, while this process is hindered when only H atoms linked to aromatic rings are available [9].

Initial states in the electropolymerization of aniline and p-aminodiphenylamine have been studied by in situ FTIR and UV-Vis spectroelectrochemistry. The measurements indicated the highest polymer growth rate during the reversed cathodic potential scan in potentiodynamic electropolymerization. In this scan, direction the radicals were formed in a less anodic potential region by synpropor-tionation of the soluble oxidized dimer [10].

Electrochemical deposition of poly(3-dodecylthiophene) was studied by the EQCM technique and an induction period was observed due to the formation of soluble oligomers. In a later stage of the oxidation, co-deposition of theses olo-gomers occurs, leading to the decrease in the mean conjugation length. This phe-nomenon was made responsible for the appearance of nonrigidity [11].

Combining Scanning Electrochemical Microscopy with Infrared Attenuated Total Reflection Spectroscopy, the access to the fingerprint region of the IR spectrum (>10 μm) not only revealed the polymerization level but provided direct spectroscopic insight on the polymerization mechanism [12]. The surface of a ZnSe ATR crystal was initially coated with 2,5-di-(2-thienyl)-pyrrole (SNS) layer, which was then locally polymerized during $Ru(bpy)_3^{2+}$ mediated feedback mode SECM experiments. The polymerization reaction was simultaneously monitored by recording absorption intensity changes of SNS specific IR bands.

A combination of EQCM and in situ FTIR spectroelectrochemisty could prove the formation of real copolymer or block polymer during the growth of polyaniline and poly(aniline-co-o-aminophenol) films [13]. The results obtained by the means of in situ piezoelectric FTIR spectroelectrochemisty indicated that the copolymer-ization process and the properties of the copolymer were different from that of polyaniline. The copolymer formed through head-to-tail coupling of the two monomers via NH-groups was a new polymer rather than a mixture of polyaniline and poly-o-aminophenol.

Morphology alteration at different stages of electrodeposition of polypyrrole in the presence of an enzyme (glucose oxidase) has been described, and formation of multilayers has been concluded by the simultaneous application of the combination of optical (ellipsometric) and electrogravimetric (EQCM) measuring techniques for the study of the electrochemical growth [14].

Mechanism of aniline oxidation, its autocatalytic pattern and the observed fluc-tuations have been obtained by simultaneous real time recording of electrochemical, spectroelectrochemical, and mass data. Observations indicated that the oxidized species of the dimers evolved into the longer polymer chain. The fluctuations in the absorbance values reflected their relative concentrations due to a series of reactions for PANI growth in which the reduced PANI is first oxidized. This process was followed by the chain-lengthening reaction, rendering part of the chain to a reduced state. This autocatalytic reaction required the oxidized and reduced states to fluctuate, which was shown to take place by the voltabsorptograms [15].

Piezoelectric diffuse reflectance spectroelectrochemistry (PDRSEC), a new technique of diffuse reflectance spectroelectrochemistry (DRSEC) in combination with electrochemical quartz crystal microbalance (EQCM), was developed to study the electrochemical copolymerization of aniline and o-anthranilic acid. The measurements proved for the poly(aniline-co-o-anthranilic acid) growth that the copolymer films were partially self-doped and more rigid than PANI. Simultaneous crystal frequency and resistance recording has provided also a deep insight into the copolymers' pH-dependent swelling/dissolution behavior [16].

4.2 Chain-Growth Process

Radical cation–monomer or cation radical—cation radical coupling

Answer to the question and longtime uncertainty whether the chain growth takes place via a coupling between two cation radicals or in a reaction of the cation radical with the monomer has been answered by UV-VIS spectroelectrochemistry. As it is well-known, during a polymerization performed in the galvanostatic mode, the deposition is continuous, which results in a gradually developing spectral increase in the whole visible region (see in Sect. 3.1). Oppositely, when polymerizing with the cyclic voltammetric technique, deposition occurs in steps or sequences, dominantly during the positive going scans, but it is interrupted by the reverse scans, turning back the film into the reduced, nonconducting state. In the subsequent cycle, first the previously deposited film is oxidized, since the redox potential of the polymer is much less positive than that of the monomer. As a consequence, the previously deposited film transforms first into the oxidized form, thus becoming able to react with the monomer molecules, available in the adjacent solution phase. However, as it was proved, no absorbance increase was detected, albeit the potential increased in this medium range, instead the subsequent identical spectra were densifying. Importantly, no further deposition was detected before the potential reached that of the monomer oxidation. From this fact, it was evident that chain-growth process did not occur in a reaction between cation radical and the thiophene monomer [17].

4.3 Nucleation and Growth Steps

An early ellipsometric study on the polymerization of thiophene from CH_3CN solutions revealed that the process is kinetically controlled. At short times there is evidence for nucleation. At longer times, nuclei overlap and uniform growth is indicated by a linear increase of film thickness with time [18].

A rapid scanning spectroscopic ellipsometer with optical multichannel detection has been employed for in situ monitoring and analysis of pyrrole electropolymerization on a gold electrode, and both dielectric function and thickness of the semitransparent polypyrrole were accurately obtained from the final thick film spectra. Three ranges could be distinguished: regimes of (i) monolayer adsorption and Au interface reordering in the first 10 Å, (ii) two-dimensional nucleation and then coalescence, which occurs at a roughly constant film thickness of 150 Å, and (iii) gradual densification, followed by layer-by-layer growth at a constant rate after 350 Å [19].

Results for a conducting polymer poly(thiophene-3-acetic acid) demonstrated that electrochemical deposition proceeded favorably through two-dimensional layer-by-layer nucleation and growth. The findings indicated the possibility to prepare ultrathin, compact conducting poly(thiophene-3-acetic acid) films [20].

4.4 Mechanistic Isuues

Proofs for the change from the bulk solution to surface coupling chain-growth mechanism

Combined rotating ring disk electrode and spectroelectrochemical evidence showed clearly that bipyrrole was formed in solution during the anodic polymerization of pyrrole in acetonitrile. An absorbance peak at 242 nm formed during electropolymerization was identified as being due to protonated pyrrole. Since this species goes on to form pyrrole black, it may influence the conductivity of the electropolymerized product [21, 22].

Since the intermediates, formed during the electrodeposition of conducting polymers, tend to interact with nucleophylic reactant, the presence of this latter has an inhibiting effect even in trace concentrations. Appropriately, addition of 1,5-diazabicyclo[4.3.0.]non-5-ene (DBN) to the solution in gradually increasing but still a trace amount resulted in the shift of the characteristic absorbance increase, indicating the changing delay of the deposition, due to the reaction of the intermediates with the inhibitor present. Beyond a critical but still paucity concentration, the film formation is totally inhibited, from which the dominancy of the side reaction between the nucleophylic partner and the soluble oligomers is concluded [23].

However, addition of the inhibitor in the same, critical concentration but *during* the electropolymerization did not effect the deposition. Thus the previously seen homogenous inhibition cannot work in this stage of the electropolymerization, which evidenced that from a certain moment the chain growth takes place on the surface in a heterogenous reaction between the oxidized film and the oxidizing monomer [17].

4.5   Regular/Irregular Behavior of the Deposited Film

The first reduction of the polymer film—UV-VIS

During the well-known reduction of polypyrrole, the midgap excitations gradually cease one after the other, and the shape of the neutral film is obtained back. This step-by-step reduction could be clearly demonstrated by voltabsorptiometric measurements [24].

However, if the reduction was performed immediately after the end of the polymerization, the absorbance of the highly oxidized species at 800 nm went through a maximum, and the midgap excitation never disappeared later: the film remained in the partially oxidized form.

A series of identically performed experiments revealed that this irregular behavior is connected to the lack of enough waiting time after the end of the polymerization. The film has to be left to relax before its first electrochemical reduction, in order to avoid irregularities and irreversible damages [25].

Consequently, it was an interesting question, what happened during the open-circuit relaxation of polymer films.

A systematic study after the end of the electropolymerization in sodium dodecylsulphate (SDS) solution demonstrated three phenomena

(i)   The increase in the conductance did not stop. As it was determined for a PPy film, it went further during the relaxation period [26].

(ii) The conductance increase was accompanied by proton removal from the layer, detected by a pH probe in the vicinity of the electrode.

(iii) This period of the process could be connected also to mass decrease, measured by EQCN. Here it is worth mentioning that—in order to get further information—laser beam deflection would be useful to detect ionic movements, since at open-circuit electroneutrality should be preserved in both the film and the solution phase.

An interpretation of the relaxation connected to these observations has also been given by concluding that still protonated segments are transforming

$$PPy^+DS^- + (HPPy)^{2+}2DS^- \rightarrow PPy_2^{2+}2DS^- + H^+_{(aq)} + DS^-_{(aq)}$$

Thus during relaxation, the transformation into the stacked state is assumed to accomplish.

As for the regular reduction, which starts with the more conjugated form $PPy_2^{2+}2A^-$, parallel formation of monocationic and neutral segments was assumed in a dissociation-type step with the participation of cations from the solution:

$$PPy_2^{2+}2DS^- + e^- + Na^+_{hydr} \rightarrow PPy^+DS^- + PPy^0 + (Na^+DS^-)_{hydr}$$

As it was experienced, the conductance increased during the relaxation period also in the case of polythiophene, and became even three times larger compared to the value measured at the end of the electrochemical polymerization.

4.6 Specific Dopant–Polymer Interactions

Electrochemically synthesized polythiophene and poly(3-substituted thiophenes)—in comparison with polypyrrole—have been characterized by visible, IR, and XPS spectroscopy, and the effect of dopant especially on the electrical properties of organic conducting polymers has been observed. Large intrachain conductivity, increasing with the doping level, was demonstrated for the thiophene derivatives, which showed a marked metallic behavior. In contrast, the relatively low macroscopic conductivity, measured on pressed pellets, was related to poor interchain contacts and to morphological inhomogenities [27].

Doping levels and the rate of absorbance changes at different potentials were determined from in situ spectroelectrochemical measurements, and the effect of spherical anions (perchlorate, tetrafluoroborate, hexafluorophosphate) on the transient redox behavior of polypyrrole in anhydrous acetonitrile has been compared. The uniformly performed experiments exhibited much larger doping level with BF_4^- which would not be foreseen, considering spherical anions of the same size. This anomaly was not evidenced for poly(N-methylpyrrole) [28], when compared with those obtained for polypyrrole [29]. In this case, BF_4^- behaved in a similar way to ClO_4^-, in contrast with polypyrrole. This observation was interpreted by that the BF_4^-—polymer interaction occurred in polypyrrole via the N-H bonds which are absent in poly(N-methylpyrrole).

Eventual covalent bond formation between a polymer segment and electronegative anion during the electropolymerization

An extraordinary cation exchange behavior evidenced by EQCN measurements has been found with poly(3,4-ethylenedioxythiophene), synthesized in the presence of chloride or fluoride ions [30].

It was surprising because chloride was always found a mobile anion. The mobility of even the bulky cetyl-$(CH_3)_3N^+$ cation was experienced when the redox transformation of the polymer was performed in the presence of hexadecyl-trimetyl ammonium chloride.

Moreover, when electropolymerization took place in the presence of halide ions, the available doping level decreased much compared to a film prepared with BF_4^-. Complementary EQCN measurements exhibited that also polypyrrole, prepared in the presence of Bu_4NCl, exhibits cation exchange behavior [31].

The phenomenon seems to be general since redox capacity of the polymer film, prepared or doped later in the presence of electronegative anions suffered from irreversible changes. The decrease in the oxidation was influenced by both the anion concentration of the doping halide anion and the oxidation potential not only during the polymerization but also during the redoping process, when a polymer—prepared in the presence of noninteracting anions—was cycled in the presence of halide ions. The irreversible interaction of both chloride and fluoride anions have been observed in the case of various polymers.

4.7 Monitoring the Formation of CP Composites

Incorporation of special anionic forms and nanoparticles

The verification of the incorporation can be based on investigation during both the deposition and the later redox studies of the given composite.

EQCN results obtained for the formation of polypyrrole/oxalate polymer film indicated the presence of the Fe(II),Fe(III)-oxalate redox active, mixed valence complex anion, formulated as $[Fe(II)Fe(III)(ox)_3]^-$ [32].

Pyrrole was oxidized in the presence of exclusively iron(II)-oxalate in its saturated solution with slurry. With assumptions for the theoretically possible doping anions (the simple oxalate^{2-}, or $[Fe(III)(ox)_3]^{3-}$, or $[Fe(II)(ox)_2]^{2-}$, or $[Fe(II)Fe(III)(ox)_3]^-$), and using the doping level calculated from the redox and the polymerization charges by the formula

$$d = 2\, Q_{ox}/(Q_{pol} - Q_{ox}),$$

the virtual molar mass of the doped monomeric units and the mass/charge ratios were calculated.

For example, if oxalate is assumed to be the dopant, the virtual molar mass M_v can be obtained from the molar mass of pyrrole ($M_r = 67$) and the divalent anion ($M_r = 88$) as well as from the doping level (0.15) as

$$M_v = (67 - 2) + 0.15 * 88/2 = 71.6$$

and for the slope

$$\Delta m/\Delta Q = M_v/(2 + 0.15)\ F$$

equation can be used.

The first three cases resulted in a slope substantially smaller than the experimental one (0.345, 0.390, and 0.397 µg/mC, respectively). Only the assumption of the mixed valence complex doping ion delivered a value larger (0.585 µg/mC) than the one obtained experimentally (0.489 µg/mC). On this basis, an assumption that the positive charges of the polymer film are at least partially compensated by the mixed valence monovalent anion was deduced. The iron(III)/iron(II) ratio in the composite could be obtained by Mössbauer spectroscopy, from which the 1:3 ratio of the mixed valence monovalent anion and $[Fe(III)(ox)_3]^{3-}$ was obtained, and the linear combination of the two doping ions resulted in a slope value close to the experimental one.

Special interaction between magnetite nanoparticles and 3-thiopheneacetic acid through the chemical bond between OH-groups—stabilizing Fe_3O_4 nanoparticles—and the carboxylic group of the monomer has been identified during the electropolymerization of magnetite incorporated polythiophene. The interaction could be exploited also to perform the layer-by-layer polymerization. By modifying the amount of iron oxide in the polymerization solution, the inorganic material content of the layer could be increased up to 80 m/m%. While electrochemical results, including data obtained by electrochemical quartz crystal microbalance (EQCM) proved that the presence of Fe_3O_4 did not influence the redox properties of the polymeric film, in the presence of magnetite an extraordinary microstructure could be detected, where the self-assembling magnetic component strongly determined the morphology of the composite, leading to aligned band formation and a microstructure with parallel strips of ~ 1 μm width [33, 34].

Since a chemical synthesis of poly(3-thiopheneacetic acid)/magnetite nanocomposites led to layers with tunable magnetic behavior, these new modified electrodes, incorporating a large amount of Fe_3O_4, may be used in magnetic electrocatalysis [35].

The composites consisting of an organic CP and an inorganic material are generally called as organic/inorganic hybrids. There sophisticated design and its realization is a key factor in synthesizing electrode with special properties. In this vain new efficace catalyst and photocatalysts can be obtained which are promising in energy-related processes. These materials can be used in both storage and generation of energy, including the exploitation of renewable energy sources. A recent review summarized the various paths together with the characterization of these hybrids by in situ techniques [36].

Detection of the incorporation of biologically active components

The simultaneous application of two in situ techniques, ellipsometric and microgravimetric (QCMB) measurements, for the study of the electrochemical growth of a conducting polypyrrole in the presence of glucose oxidase provided information on fundamental properties of the enzyme-containing film, including film thickness, mass, and density. The results showed that incorporation of the enzyme resulted in changes in the apparent optical properties and in the apparent density of the electrochemically grown film, and suggested the mutual stabilization of the polypyrrole and the enzyme in the composite layer [14].

PPy/magnetite/vitaminB12 composites have been synthesized and studied by EQCN electrogravimetry and the incorporation of both the magnetic nanoparticles and the biomolecules were demonstrated. The magnetite nanoparticles were soaked in the enzyme solutions, and adsorption of this latter could be demonstrated by the decrease in the characteristic absorbance intensity of the enzyme. The linear mass increase of the layers versus the transferred charge during the synthesis under totally identical electrochemical conditions for the neat polypyrrole, the film polymerized in the presence of magnetite, and the film prepared with magnetite, covered with adsorbed B12 were compared. From the slope values, the relative amount of the built-in materials could be calculated, for which 26.7 m/m% magnetite and 15.0 m/m% B12 were obtained [37]. Similar experiments could be repeated for the PPy/magnetite/laccaze film.

Electrochemical quartz crystal microbalance (EQCM) was applied to monitor the process of immobilization and release of anticancer drug, disuccinyl derivative of betulin, in PEDOT matrix. Subsequent steps of the process have been investigated, i.e., electrochemical polymerization of monomer in the absence of drug, removal of primary dopant during the process of matrix reduction and drug incorporation during the process of matrix oxidation. Furthermore, the release of drug from PEDOT matrix has been performed and followed during the spontaneous release with no application of external potential [38].

Signal enhancement in sensor applications

Implementation of electrochemistry with plasmonic nanostructures for combined electrical and optical signal transduction has been used in [39]. A common ESPR sensor for glucose sensing based on conductive polymer films with the enzyme glucose oxidase incorporated operated by monitoring the current from enzymatic oxidation at a maintained potential. The electrochromic properties of the surrounding polymer matrix provided an RI contrast upon glucose oxidation which was high enough for the SPR readout to provide better signal to noise than ordinary chronoamperometry.

4.8 Details of the Redox Switching

The first but very fundamental summary of in situ UV-VIS spectroelectrochemical observations has been given in an early review [40].

In this review, the authors summarized the optical studies of some of the key conducting organic polymers that have become important by that time, and proof that the optical properties of conducting polymers are important to the development of an understanding of the basic electronic structure of the material. The in situ technique described for studying the charge-transfer doping reactions in conducting polymers has made possible to determine the nature of the charge-storage states as well as to monitor the kinetics of the charge-transfer reaction. It was the first evidence that in conjugated polymers other than polyacetylene, electrons added or removed from the delocalized π-bonded backbone initially produce polarons (radical ions coupled to a spatially extended distortion of the bond lengths) which subsequently combine to form dianions or dications (spinless bipolarons), respectively. In polyacetylene anions and cations but not radical anions or radical cations were produced during charge transfer (spinless negative or positive solitons or charged bond-alternation domain walls). The authors laid down that the high-contrast electrochromic phenomenon associated with electrochemical doping and the characteristic changes in absorption spectra appear to be general features of conducting polymers. The oscillator strength associated with the interband transition prior to doping shifts into the free carrier contribution in the infrared after doping. The effect of such spectral changes depends initially on the magnitude of the energy gap. If E_g is greater than 3 eV, the undoped insulating polymer is transparent (or lightly colored), whereas after doping the conducting polymer is typically highly absorbing in the visible. If, however, E_g is small ($\sim$ 1–1.5 eV), the undoped polymer will be highly absorbing, whereas after doping the free carrier absorption can be relatively weak in the visible, rendering the polymers transparent in this visible region of the electromagnetic spectrum. Another result of fundamental importance described was the discovery that polymers that are soluble (and hence processible) in common organic solvents and in water show the same spectroscopic characteristics in dilute solution as in the solid state, particularly as a function of doping; i.e., charge storage in the form of dications is not relegated to a solid state effect but is an intrinsic, single-molecule effect.

The essential feature required for conducting polymers is conjugated systems with a π-electron band structure. Such polymers can undergo charge-transfer reactions electrochemically or with suitable electron donors or acceptors to provide potential carriers. The conjugation is related to the coplanarity along the heteroaromatic ring system; a conclusion confirmed with thiophene oligomers [41].

As an interpretation of the spinless charge carriers in nondegenerate polymers, the idea of "π-stack" by Miller [42] or of reversibly decoupling "σ-bond" formation by Heinze [43] between paramagnetic species was later also assumed.

The influence of alkyl side chain distribution patterns on the spectroelectrochemical properties of poly(alkylthiophenes) has been studied by Pron et al. The two types of compounds were of different coupling pattern: poly(4,4'-dialkyl-2, 2'-bithiophenes), which represented head-to-head and tail-to-tail coupled poly (alkylthiophene) chains, and poly(3-alkylthiophenes), which represented head-to-tail coupled poly(alkylthiophene) chains.

Both types of polymers could be obtained chemically or electrochemically by the constant potential method or the potential scanning method. Poly(4,4'-dialkyl-2,2'-bithiophenes) differed significantly from poly(3-alkylthiophenes) in their voltammetric and spectroelectrochemical properties. The oxidative doping of these compounds occurred in a very narrow potential range and was significantly retarded in dynamic measurements. Analysis of the changes in absorption in the 500 nm spectral region and the observation of an induction period in the oxidation at constant potential indicated that the oxidative doping was preceded by structural changes, interpreted by a faradaic phase transition [44].

In situ ellipsometry and Fourier Transform Infra Red spectroscopy data on the growth, electrochemical cycling and overoxidation of polypyrrole showed that the initial charge-carrying species were polarons. At higher potentials, bipolarons were also formed, and coexisted with the polarons, strongly suggesting that there was no appreciable energy gain in forming a bipolaron (with respect to two polarons). Both species have narrow, well-defined conjugation lengths, with bipolarons being ca. 9 monomer units smaller than polarons. The oxidation of the polypyrrole resulted in the expulsion of solvated protons from the film. In addition, oxidation of the film was accompanied by a reversible 30% reduction in its thickness [45].

The formation and stabilization of charge carriers in polyaniline during p-doping was followed in dependence of the chain branching [46].

The potential dependence of the IR bands during the oxidation of the polymer clearly demonstrated the formation of the different charged polymer structures such as π-dimers, polarons, and bipolarons. It has been shown that IR bands usually attributed to a semiquinoid polaron lattice corresponded in fact to doubly charged species, π-dimers, which are face-to-face complexes of two polarons. Bands corresponding exclusively to polarons have been identified at 1266, 1033, and 1010 cm^{-1}, suggesting that polarons are predominantly stabilized on the linear segments near the polymer branches by phenazine. The multicomponent charged state and their interconversion was made responsible for the hysteresis.

A method for electrochemically in situ conductivity measurements based on a glassy carbon fiber array double electrode was described [47].

The electroactivity and potential dependent conductivity of polypyrrole film were strongly affected by solvent and the doping anion's solubility in the solvent, and also by the history of electrochemical treatments in different electrolyte solutions. It was an interesting observation that NO_3-doped polypyrrole could completely keep its conducting state (doped state) at a reasonably negative potential (e.g., −0.8 V vs. SCE) in acetonitrile solutions.

The redox transition of the poly(o-aminophenol) from its completely oxidized state to its completely reduced state occurred through two consecutive reactions in which a charged intermediate species took part. UV-Vis and Raman signals agreed with an increase of the concentration of an intermediate species until the potential of the maximum redox peak, which later diminished with the potential. The results of in situ FTIR spectroscopy agreed with Raman measurements [48].

In situ surface-enhanced resonance Raman spectroscopy (SERRS) with excitation at 1064 nm has been used to monitor the dynamic changes in charge transfer, structure, and orientation during the redox doping of a self-assembled monolayer of conducting poly-3-(3'-thienyloxy)propanesulfonate (P3TOPS). The SERRS spectra were compared with Fourier transform Raman spectra of solid 3TOPS monomer and with reduced and oxidized forms of P3TOPS in solution. Data indicated that the oxidation process went through initial preferential generation of polaron charge carriers with followed generation of bipolarons at higher oxidation levels. At low doping levels, thiophene rings were more coplanar than at higher ones. At higher potentials applied, the trans → gauche conformational transformation of the alkoxy side chain was observed along with the massive growing number of bipolarons and simultaneous deplanarization of the polymer chain [49].

The optical beam deflection technique (mirage effect) showed unambiguously that the ionic compensation of positive charges created in polyphenylene films on an electrode was affected only by the anion during the whole oxidation process. Use of a new technique of convolution allowed quantitative determination of the amount of exchanged species during cyclic voltammetry, moreover the correction of the propagation delay between the electrode and the laser probe became also possible with this convolutive approach [50].

It has been demonstrated soon after that the optical signal is generated not only by surface-transferred ionic species but also by a redistribution of background ions, and appropriate choice of the background electrolyte is needed to avoid these complications [51].

The optical beam deflection technique has been used simultaneously also during cyclic voltammetry and chronoamperometry on two electrodeposited thin films in the presence of an aqueous solution of lithium perchlorate: polypyrrole and the poly {pyrrole-co-[4-(pyrrole-1-yl)-butane sulfonate]} which latter contains negatively charged ionic groups attached to the polymeric chains. The beam deflection allowed a good identification of either the uniqueness of the ion exchange or of the multiplicity along the whole potential region [52].

Parallel application of the in situ conductivity and ESR revealed a very special correlation. In situ conductivity of polypyrrole (as tosylate) as a function of oxidative doping level attained a maximum at 75% of the total oxidation charge, and the relevant in situ ESR signal corresponded there to an equal concentration of spin-carrying (polaron) and spinless (bipolaron) species. Results were explained on the basis of mixed-valence conduction [53].

Identification of the components of the total charge

An important part of studies on the details of the redox transformation was connected to the separation of the charge components and their assignment according to different types.

The capacitive versus non-capacitive charge in conducting polymer electrodes was first discussed, when anomalously large current plateaus were observed in the cyclic voltammograms of conducting polymers such as polypyrrole, and have been interpreted as the charging of a large capacitance attained by the polymer in the oxidized state, and can be described in terms of an ionic relaxation mechanism. Analyzing the nature of the changes stored in chemically synthesized polypyrrole by a.c. impedance and cyclic voltammetry, the total current has been decomposed into two components: a capacitive current and a non-capacitive one [54].

Impedance spectroscopy was used to study the redox mechanism of poly-3-methylthiophene. It has been demonstrated that the redox process involved two types of doping. A fast doping occurred at the surface of aggregates of compact chains and it did not require the penetration of counterions between the chains. This type of doping had a large capacitive effect. The second type of doping was a bulk phenomenon being equally important as the first, but its kinetic was limited by the diffusion of counterions between the chains and, as a result, it appeared slower than the capacitive phenomenon. A model of the redox process was drawn up for reduction where two reduction steps were considered with two different redox potentials, that would correspond either to two types of oxidizable sites in the material or to two electronic steps. For both of them, the total charge was separated in a fast capacitive component and a slow faradaic component, and a relation connecting the two processes was derived [55].

Spectral changes in the visible wavelength range were studied versus the transferred charge during the redox switching of poly(3-methylthiophene) [56].

The measurements proved that the modification of the visible spectrum comes to an end during the period connected to the half of the transferred total charge. Although further changes may—and do definitely—take place in the near IR region, the second part of the charge was not coupled with the decrease in the wavelength range below ~ 500 nm, excluding the further oxidation of the neutral segments in this stage of the process.

The charge–potential relationship during the oxidation proved linear in the second section of optically constant pattern, which allowed assuming a capacitive behavior of constant capacitance.

Its slope led to an enormously large capacitance value which is related to the use of the "supercapacitor" term for conducting polymers. However, this capacitance is connected to the redox transformation of the film hence it should be distinguished from double layer capacitance.

During the reduction, first this capacitance seemed to be discharged, and the return of the spectrum to the original, neutral form took place delayed, only during the second half of the total charge—giving a rational background of the generally reported hysteresis.

The oxidation peaks for the redox switching of electrochemically prepared polyaniline (PANI) films were studied in H_2SO_4. The absorption coefficient of the polaron has been derived from the relationship between the modulated transmittance and the film capacitance which also showed that bipolaron states were formed at potentials more positive than 0.15 V versus SCE [57].

The combination of optical and electrical measurements allowed Faradaic and non-Faradaic charging processes to be distinguished. The analysis of the modulated transmission data indicated that the capacitance measured in a.c. experiments is associated with redox transformations in the PANI film. The a.c. capacitance was found to be lower than the pseudo-capacitance derived from cyclic voltammograms.

Based on some reports, the possibility of a correlation between the different types of the charge and morphological changes has assumed.

In situ ellipsometric study of the growth and electrochemical cycling of poly-pyrrole films from the reduced to the oxidized form revealed that the film thickness decreased between ca. −0.6 and −0.2 V, but thereafter, although charge was still passed, it changed much more slowly [58].

Upon reduction (undoping) in aqueous solution of $NaClO_4$, a perchlorate-doped PPy underwent initially rapid swelling and subsequent continuous slow shrinking during the first reduction potential step from 0.2 to 0.8 V, but during subsequent redox potential steps no remarkable changes of volume were observed. In contrast, in an aqueous solution of sodium p-toluenesulfonate, the polymer swelled on reduction and shrinked on oxidation during continuous redox potential steps [59].

A different electrochemical behavior of polypyrrole in different solvents was observed. Differences were attributed to the presence of interactions between solvent and polymer. These interactions changed when polymer was oxidized or reduced [60]. Since the high polarity of water promotes an intense interaction with polarons and bipolarons, counterions and water molecules penetrate in the polymer during oxidation, promoting an important swelling and molecular strain. This phenomenon can be utilized to provoke movements using polymer bilayers [61].

Observation of hysteresis

Cyclic voltammetric studies on the redox transformation of the majority of CPs exhibited a well identifiable hysteresis: a generally single but broad oxidation wave, and a usually two-step reduction. Under appropriate conditions, the total charge obtained by integration is zero, indicating the global reversibility of the process.

The hysteresis, observed on the cyclic voltammetric curves during the redox transformations, was revealed by numerous kinds of the in situ techniques. Even in the simplest cases, when only easily mobile anions are participating in the charge compensation process, the mass–potential curve, and also the mass–charge curve, obtained by EQCM, exhibit a loop, moreover, the mass of the film is not the same as it was before the scan, (it is generally smaller), and this lack of mass is recovered during waiting at the starting potential or even at open-circuit.

The rate of the mass change curves—obtained as the derivative of the mass change along time—had a similar shape as the voltammogram, exhibiting the existence of the hysteresis.

The hysteresis and the delay of the setting of the conducting state could be well illustrated using various scanning rates. The faster the potential was changed, the smaller conductance value was achieved at the positive potentials, with a maximum conductance perceived always on the backward region [62].

Importantly, the only plot exempt of hysteresis was the conductance versus the logarithm of charge! This fact is reminiscent to the formula in Sect. 3.1.6, used for the determination of the conductivity during the electrodeposition.

The hysteresis phenomenon was interpreted by the scheme-of-cube model by Hillman [63].

The model considered electron/counterion transfer, solvent transfer and polymer reconfiguration as distinguishable processes. The participation of solvent transfer and polymer reconfiguration was found to depend on the experimental timescale, while the initial polymer film state was demonstrated to be determined by the film's history.

4.9 Development of the Conductance

As it has been demonstrated, the development of the conducting state of polythiophene can be inhibited or postponed by a negative pre-polarization without a Faradaic process [64].

Voltammetric cycles were started from −0.4 V to the cathodic direction till the negative turn-points to −1.0, −1.3, −1.5, −1.8, and −2.0 V, respectively. Then they were running to the positive range to oxidize the film. The cathodic polarization—

although in a range where no Faraday-type current is detected—inhibited the development of the conducting state to various extent, and a gradually smaller conductance could be achieved correlating with the negative end value of the pre-polarization, under otherwise identical experimental circumstances. Due to the fact that no cathodic doping occurred, the phenomenon can be connected assumingly to solvation/morphology changes.

In situ conductance measurements with poly(3-methylthiophene) film in Bu$_4$NPF$_6$/acetonitrile base solution provided direct evidence for that—after the oxidation of the film—the accomplishment of the state of high conduction is not necessarily coupled with current flow any more. The non-faradaic step which may explain this observation was described as a chemical reorganization within the film, which involves the transformation of solvated polymer segments to a more conjugated, desolvated form.

Furthermore, the effect of the interruption of the CV at various oxidative potentials was also studied in the way that the potential was held at the actual values till the current dropped to a near zero value. These interruptions were performed on either the anodic or the cathodic branch of the sweep. During the current relaxation the conductance always increased. The correlation between the conductance change and the potential of the interruption indicated that the tendency of the film towards turning into the conducting form depended on the "waiting" value of the potential, irrespective of the direction of the actual potential change [62].

Conductance can increase even during a negative potential jump

Multiple double potential steps—starting from the reduced state (−0.4 V) to the oxidized state (+1.3 V) and backwards to different potentials, related to the reduction of the film to a various extent—have been applied for poly (3-methylthiophene). In the case of intense reduction steps, the conductance of the film decreased proportionately. However, during a potential jump from 1.3 V back to 1.1 V, the prompt conduction decrease turned back and started to increase again at this constant potential value [65].

The nonzero remaining current beyond the oxidation peak potential in Ref. [54] was reproduced and the voltammograms were analyzed also at various sweep rates. When the values of the peak potentials for both the anodic and cathodic parts were determined, a surprising phenomenon was observed: while at faster scanning rates the anodic peak is "normally" situated, at an about 50 mV more positive potential compared to the cathodic peak, at low rates the position of the two peaks is totally irregular. The potential of the anodic peak became shifted to a value by 70 mV more negative than the value of the cathodic peak. Thus, the two peaks are not coupled to each other as a common redox pair. As the strange phenomenon developed with the decrease in the scan rate, it was assumed that some time-consuming transformation took place as far as the sweep rate made it possible. This slow transformation might be coupled certainly with the nonzero current beyond the anodic peak, and at small scan rates it had enough time to take place.

4.10   p-Type and n-Type Doping

Observation and approval of the cathodic or n-doping of CPs

The phenomenon was first reported for polythiophenes when the film could be reduced in strictly anhydrous media [66].

The electrochemical behavior of n-doped poly(2,2′-bithiophene) was studied and characterized in acetonitrile. The results showed that doping was strongly dependent both on the solvent's properties and on the size of counterion. It can be assumed that water traces undergo to a reductive transformation, leading to irreversible reaction with the polymer, while the actual cation's size is an important factor in relation with its fitting the pore and channel size of the film.

In the late 1980s and early 1990s the interpretation of the redox transformation was in the center of the research [1–5, 42, 43]. In this respect, the question of the isosbestic points on the UV-Vis spectra during the redox transformation became the topic of a scientific debate. The existence of a single well-defined isosbestic point would be a proof for the interconversion of two species, so it would have implied that only one charge carrier formed. When studying this phenomenon, the appearance of even more than one isosbestic points was reported, concluded from the saddle points on the contour plot of the absorbance spectra [67].

Further control measurements revealed that the third isosbestic point in the more negative potential region belongs to the n-doping process, but the hypothesis of the existence of one single interconversion, the interpretation of the redox transformation, involving only two species, became once and for all excluded.

Spectroelectrochemical investigation of the reductive transformation of polythiophenes illustrated that a cathodic charge alone—even with its seemingly reversible counterpart on the backward section—is not an adequate evidence for that an n-doped state was achieved. Instead, in situ taken spectra during the reduction of PTh between −0.4 and −1.8 V showed reversible optical changes—rather similar to those registered during its p-doping—serving as a real proof for the realization of the cathodic doping.

In situ EPR spectra for both p- and n-doped poly(benzo[c]thiophene) (PBCT) thin films are reported, and the spin concentration showed a maximum with doping in either sense. This was anticipated from spectroelectrochemical data for the p-doped but not the n-doped polymer [68].

Spectroelectrochemical studies of polydithienothiophenes, low band gap conjugated polymers with polythiophene-like chain were performed, using in situ FTIR-ATR and ESR spectroscopy, and the formation of paramagnetic positive and negative charge carriers with unusually high g-values were described [69].

A two-step process can be generally revealed during the return from the doped state

A detailed analysis of the dedoping of polyterthiophenes revealed the separation of two, spectrally differently behaving sections during the return from the p-doped state. The first period of the reversed transformation causes absorbance increase in the range of both the monocationic and neutral species, indicating a dissociation-type step, followed by the second part of the film discharge, leading to the recovery of the totally non-conducting form [70].

This behavior could be demonstrated also during the reoxidation from the cathodically doped conducting state. In the first region, absorbance increased in both ranges of the spectrum, indicating a dissociation-type step, similarly to the case of anodic doping. During the second step spectra return to the original form. Similar behavior of the parent polythiophene has been evidenced during its cathodic redox transformation. Thus it has been proved that redox transformations of the neutral, non-conducting polymer are symmetrical processes, and p-type and n-type forms have the analogous excitation characters. The optical behavior of both charged state is the same, independently whether their generation was performed by electron withdrawal (p-type) or electron addition (n-type) [71].

In this study also the shift in the absorbance maximum of the neutral form during the redox transformation was discussed. A blue shift of the absorbance maximum for the neutral species during the oxidation has been generally connected to the gradually decreasing effective conjugation length of the actually transformed polymer segments, indirectly meaning that the red shift during an opposite polarization is based on the same reasoning. This conclusion has been question-marked when the blue shift was monitored during the undoping process.

Neutral poly(3,4-ethylenedioxythiophene) (PEDOT) thin films can also be switched to an electronically conducting form either by oxidation (p-doping) or reduction (n-doping) in anhydrous organic solvents [72].

The maximum attainable n-conductivity was ca. 1% of the maximum p-conductivity. However, based on spectroelectrochemical and in situ conductance measurements, the p-conductivity regime could be divided into two domains, in which either positive polarons or bipolarons and free carriers were the major charge carriers. In the n-conductivity regime, voltammetric, spectral, and conductance data suggested only the generation of negative polaron-type carriers. The results implied that the conductivity due to positive or negative polarons is of the same order of magnitude and that the higher maximum p-conductivity may be attributed to the generation of other charge carriers in the highly stable oxidized PEDOT films.

A review presented an overview of the trends of the application of p-conjugated oligomers and polymers in field-effect transistors, light-emitting diodes, electrochromic devices, and solar cells. The area has introduced major changes in the chemistry of gap engineering, and the paper summarized the electrochemical and optical properties of several conjugated fluorophores together with their sophisticated design [73].

As it was emphasized, active materials for electronic and photonic applications must present appropriate absorption and/or emission properties, highest occupied and lowest unoccupied molecular orbital (HOMO and LUMO) energy levels and charge-transport properties.

In situ FTIR spectroscopy, and d.c. and a.c. electrical measurements were used to characterize an electronically conducting polymer prepared from the electropolymerization of a bridged dithienyl monomer with low band gap energy [74].

Poly(dithienothiophene)s (PDTTs), low band gap conjugated polymers with polythiophene-like chain were studied using Raman spectroscopy, photoinduced infrared absorption, as well as attenuated total reflection Fourier transform infrared (ATR-FTIR) and electron spin resonance (ESR) spectroelectrochemistry. The spectroelectrochemical studies were performed in situ during p- and n-doping. By means of in situ ESR spectroscopy, the formation of paramagnetic positive and negative charge carriers with unusually high g-factors could be proved [75].

The modifications in the electronic distribution upon electrochemical p-doping (oxidation) and n-doping (reduction) of poly(p-phenylenevinylene) (PPV) film have been studied in situ by resonance Raman spectroscopy, optical absorption spectroscopy and ESR spectroscopy [76].

Results led to the conclusion that charge transfer in this system was mainly accomplished by polaron species formed upon doping of the polymer. In this reaction the quinoid structure was formed rather than the benzenoid structure.

Determination of the band gap energy

There is a distinction between "optical band gap" and "electrical band gap" (or "transport gap"). The optical band gap is the threshold for photons to be absorbed, while the transport gap is the threshold for creating an electron–hole pair that is not bound together. The optical band gap is at a lower energy than the transport gap. For approximate calculation, the potentials required for the onset of the p- and n-doping, respectively, are needed, and their difference gives the energy of the band gap. Electronic absorption bands were observed during both n- and p-doping processes. The difference in the onset potential of these processes can be used to calculate the energy of the band gap [70, 71].

The 2.2 eV band gap of polythiophene can be obtained from the onset potential values 0.9 V for the anodic and −1.3 V for the cathodic doping, respectively, and the difference is 2.2 V, equivalent to 2.2 eV for a one electron transfer.

According to the other method, a rough guess for the band gap energy is based on the optical spectrum of the neutral polymer, since

$$E = \mathrm{h} \cdot \mathrm{c} / \lambda$$

The absorbance peak at the larger wavelength side is extrapolated to the wavelength axes to get the highest wavelength value λ at which the semiconductor can be excited. The calculation using the Planck constant, the speed of light and the ~ 630 nm limiting wavelength in Ref. [71] results ~ 1.97 eV.

A light-emitting diode (LED) is a p–n junction diode, which emits light when activated. Applying a suitable voltage, electrons are able to recombine with electron holes within the device, releasing energy in the form of photons. This effect is called electroluminescence, and the color of the light (corresponding to the energy of the photon) is determined by the band gap energy of the semiconductor.

Conducting polymers from ambipolar monomers

Recently, a new group of alternating donor–acceptor compounds received special interest of the field: this series of new π–conjugated polymers contains both electron donor and electron accepting units. Electron acceptor groups (derivatives of thiazole, oxazole, imidazoles, triazines, tetrazine) combined with electron donor polymers allow to obtain new conjugated polymers revealing interesting absorption bands in the medium or low-energy part of the spectral range of visible light. These alternative donor–acceptor architectures make possible to prepare polymers able to be charged both positively and negatively. The injected positive and negative charges are distributed over the whole polymer backbone, thus an ambipolar D-A-type polymer is capable of conducting both holes and electrons. Since this ambipolar behavior is connected to different moieties along the chain, these materials open the opportunity for designing conducting polymers with tunable bang gap energy and electronic conductivity. For this reason, they can be applied in organic optoelectronics and electrochromic devices, being an attractive alternative to traditional crystalline inorganic semiconductors with new perspectives in organic light-emitting (OLED) and photovoltaic devices.

For reference, some important contributions are selected [77–89].

4.11 Identification of the Charge Carriers

Due to the overwhelming works in this respect, a restricted number of results—selected arbitrary from the most cited items—are summarized in chronology with the main purpose to illustrate the enormous efforts dedicated to this issue.

Midgap energy states—called polarons and bipolarons—have been identified very early during the conducting polymer story [1–5, 90].

For poly(2-methylaniline) it has been found that a protonated (bipolaronic) form can be stabilized in the fully oxidized polymer at pH <-0.5 and this new electronic structure can be characterized by an absorption band at 630 nm [91].

However, in situ conductivity measurements have shown an insulating behavior for this protonated fully oxidized form whereas the partially oxidized (polaronic) structure is conducting.

In the case of CPs with non-degenerate ground state the aromatic—quinoidal shifts during the oxidation/reduction processes have been detected by resonance Raman scattering for both polyparaphenylene and polyphenylene-vinylene at different excitation wavelengths [92].

A series of the absorption spectrum changes in polypyrrole films were analyzed using the Nernst equation by two models: the "monomer unit model" and the "polaron/bipolaron model" [93].

Formal electrode potentials and the number of involved electron were obtained and used to fit calculated ΔAbs versus E curves to the experimental plots at three wavelengths. In the second model, the "apparent" excess chemical potentials were introduced to correct for the large deviations between the Nernst equation and the Nernst plots obtained from the absorption spectra. Advantages of the monomer unit model over the polaron/bipolaron model were pointed out in the precise analyses of the spectroelectrochemical behavior of the films.

Results from in situ ellipsometry and Fourier Transform Infra Red spectroscopy on the growth, electrochemical cycling and overoxidation of polypyrrole showed that the initial charge-carrying species are polarons. At higher potentials, bipolarons were also formed, and coexisted with the polarons [45].

Anodically and cathodically conducting poly(3-methylthiophene) have been studied by transient and steady-state spectroelectrochemical methods [94].

Spectral differences were interpreted by assuming segments of different effective conjugation lengths in the neutral polymer determined by an interaction between the anions and the chains during the electrochemical preparation of the film. A redox mechanism based on thermodynamic considerations has been suggested where both the anodic and cathodic doping start with the electrochemical transformation of species of longer effective conjugation length, which leads to dications. Cyclic measurements have revealed nonequilibrium or quasi-reversible effects explained by the assumption of a two-phase system resulting in the undoping not being the simple inverse process of doping.

Factor analysis enabled to use all the information contained in the spectra, a considerable improvement to former techniques employing data at only some wavelengths [95].

The essence of the method is the application of factor analysis to the spectra recorded during a voltammetric measurement. When applied to poly(3-methylthiophene), two successive oxidation processes could be observed during the single oxidation peak in the cyclic voltammogram. The usefulness of the method in cases where voltammetric measurements are difficult to interpret was also demonstrated.

Via a combination of open-circuit relaxation and UV-VIS spectroelectrochemical measurements has been proved that radical cations (polarons)—dications (bipolarons) carrier evolution pathway critically depends on the O_2 content of the ambient in which neutral polypyrrole is electrochemically or chemically conditioned [96].

Polaron state is unstable in polypyrrole in O_2, and the polymer undergoes a direct $PP^0 \rightarrow PP^{2+}$ oxidation in this case. On the other hand, oxidation in N_2 involves the usual $PP^0 \rightarrow PP^+ \rightarrow PP^{2+}$ sequence. The poor resolution typical of cyclic voltammetric data on polypyrrole was underlined when UV-VIS spectroelectrochemical data were analyzed in the form of dA/dt versus potential plots. The utility of Raman scattering data was also illustrated via the use of change of the rate in the intensity of scattered light versus potential plots.

The spin concentration obtained from in situ EPR spectra for both p- and n-doped poly(benzo[c]thiophene) thin films showed a maximum with doping in either sense, from which stabilization of spinless charge carriers was concluded [68].

Spectroelectrochemical behavior of regioregular poly(3-octylthiophene) has been investigated using UV-VIS-NIR and Raman spectroscopies [97].

Static and dynamic UV-VIS-NIR spectroelectrochemical experiments combined with cyclic voltammetry showed that oxidative doping of the regioregular polymer is a two-step process in which polarons (radical cations) are first created which then recombine to bipolarons (dications). This two-step oxidative doping mechanism was corroborated by FT Raman spectroelectrochemical studies which showed significant changes in the positions and intensities of the Raman bands coinciding with the first and second oxidation peaks in cyclic voltammetry. These changes could be interpreted in terms of the doping induced formation of quinoid sequence of bands in the oxidized polymer. Vibrational calculations carried out for undoped and doped poly(3-octylthiophene) gave a very good agreement between the calculated Raman band frequencies and those recorded experimentally for the regioregular polymer.

The existence of indamine cation radicals in the anodic oxidation of aniline in acidified dimethylsulfoxide (DMSO) was proved by in situ and ex situ ESR spectroscopy [98].

The occurrence of radicals formed in DMSO during electrochemical oxidation was in agreement with the mechanism of the initial stages of electropolymerization of aniline in aqueous solution found by several authors.

The redox behavior of polypyrrole layers prepared by the electrochemical oxidation of pyrrole in acetonitrile solutions was studied using in situ EPR/UV-VIS spectroelectrochemistry [99].

Quantitative time dependences of the three oxidation states in the polymer layer were obtained during redox cycling and the different mechanisms of the polymer oxidation, the polaron/bipolaron route and polaron/dimer route, were compared.

By simultaneous use of ESR and UV-Vis spectroscopy the separation of the ultraviolet-visible spectra of the redox states of further conducting polymers could be achieved [100].

The electrochemical oxidation of polyaniline films on gold electrodes was studied by simultaneous electron spin resonance (ESR) and UV-vis spectroelectrochemistry. The technique permitted the separation of the superimposed UV-vis spectra of three redox states: the protonated reduced polyaniline chains, the deprotonated polaronic state, and the bipolaronic state. The separated UV-vis spectra as well as the potential dependence of each redox state concentration enabled a more precise determination of the energy levels and the stability of paramagnetic species.

Electrochemical oxidation of poly(3,4-ethylenedioxythiophene) (PEDOT) was investigated by cyclic voltammetry combined with in situ conductivity measurements, UV-vis-NIR and Raman spectroelectrochemical studies [101].

The results indicated that the doping can be adequately described by a heterogeneous "two-phase" model in which the doped conducting phase grows on the expense of the undoped, neutral phase. Doping induced changes in Raman spectra of PEDOT indicated that—unlike in other polythiophenes—the ground state of the neutral polymer is quinoid in nature whereas upon doping it is transformed into the benzenoid one.

Neutral poly(3,4-ethylenedioxythiophene) (PEDOT) thin films were switched to an electronically conducting form either by oxidation (p-doping) or reduction (n-doping) [72].

The p-conductivity regime could be divided into two domains, in which either positive polarons or bipolarons and free carriers are the major charge carriers. In the n-conductivity regime, voltammetric, spectral, and conductance data suggest only the generation of negative polaron-type carriers.

Electrochemical p-doping (oxidation) and n-doping (reduction) of poly (p-phenylenevinylene) (PPV) film have been studied in situ by resonance Raman spectroscopy, optical absorption spectroscopy and ESR spectroscopy. The Raman spectrum for electrochemically polymerized PPV was compared to infrared-active vibration bands for electrochemically n-doped PPV. When the polymer undergoes redox reactions, shifts and broadening of Raman bands, compared to neutral PPV, were observed. Interpretation of the Raman spectra and the ESR results led to the conclusion that charge transfer in this system is mainly accomplished by polaron species formed upon doping of the polymer. In this reaction the quinoid structure is formed rather than the benzenoid structure [76].

Poly(dithienothiophene)s (PDTTs), low band gap conjugated polymers with polythiophene-like chain, where an aromatic thienothiophene moiety is fused to each thiophene ring, were studied using Raman spectroscopy, photoinduced infrared absorption, as well as attenuated total reflection Fourier transform infrared (ATR-FTIR) and electron spin resonance (ESR) spectroelectrochemistry [75].

The spectroelectrochemical studies were performed in situ during p- and n-doping. Raman lines of the pristine polymers were compared to infrared-active vibration bands due to the charge carriers injected by the electrochemical doping processes or by illumination. The different π-electron distribution along the polythiophene-like chain, which determined the different band gap sizes, also accounted for the different lattice relaxations and vibrational behaviors shown by these polymers. By means of in situ ESR spectroscopy, the formation of paramagnetic positive and negative charge carriers with unusually high g-factors could be proved.

Evidence has been presented that the cycling of PPy films up to 4.6–4.8 V is reversible for the first few cycles [102].

In such a wide potential window, the maximum reversible charge obtained during the PPy film's reduction corresponded to one electron per pyrrole ring. In parallel, irreversible oxidation took place because the evolving dications located on the polymeric chains might be subjected to nucleophilic attacks.

Poly(3,4-ethylenedioxythiophene) and poly(3,4-butylenedioxythiophene) films were studied by in situ EPR spectroelectrochemistry [103].

These polymers displayed distinct narrow EPR lines at the end of the reduction half-cycle suggesting that a noteworthy concentration of spins existed in them even in the fully reduced (dedoped) state.

Radical cationic structures (polarons) were found to primarily form upon oxidation in chains of high conjugation length. However, they fully disproportionate into neutral and dicationic segments and the spinless charge carriers dominate [104].

Equilibrium and kinetics were compared for conjugated polymers, supposing the charged states were either polarons (Ps) and bipolarons (BPs) or polarons and polaron pairs (PPs) [105].

It has been described that for low concentrations equilibrium and kinetics are virtually indistinguishable. Both are essentially different for high concentrations since the extension of the bipolaron is almost the same as that of one polaron (P) whereas the polaron pair has the extension of two polarons. In contrast, for the system with PPs both P and PP concentrations saturate at high potentials. Thus, the PP model is only reasonable if an additional subsequent second oxidation step is possible.

The in situ recording of the conductance changes during cyclic voltammetry measurements (electropolymerization and the subsequent doping) of N-methylaniline (NMA) were reported [106].

The vibrational spectra of poly(N-methylaniline) (PNMA) showed specific and characteristic features not only related to its chemical structure, but also to the existence of free charge carriers delocalized along the pi-electron network.

Simultaneous electron spin resonance (ESR) and UV-vis-NIR measurements have been carried out during oxidation of a poly [N-methyl(aniline)] (PNMA) film. In situ conductance measurements were also performed during the electrochemical doping of PNMA, providing the correlation between the spin and spinless species and the conductance of PNMA during the electrochemical doping in acidic aqueous solutions. PNMA differed from other conjugated polymers (e.g., polypyrrole) in that the results from ESR and in situ conductance were consistent. At higher doping (oxidation) levels, the conductance decreased with the decrease in the ESR signal, indicating that there was no spinless bipolaronic conducting state in PNMA. Polaron pairs were supposed to be responsible for these decreases [107].

The electrochemical oxidation of polyaniline films on gold electrodes was studied by simultaneous electron spin resonance and UV-vis-NIR spectroelectrochemistry. The technique permitted the separation of the superimposed UV-vis spectra of three redox states: the protonated reduced polyaniline chains, the deprotonated polaronic state, and the bipolaronic state [108].

In situ ESR-UV/vis-NIR spectroelectrochemistry of polyaniline and its copolymer pointed to preferred stabilization of a polaron pair in the charged states at the initial charge-transfer reaction instead of polarons that were formed by equilibrium reaction at higher electrode potentials. The potential dependence of the polaron formation in PANI and the phenazine copolymer showed a potential difference between the half-peak current and the ESR intensity (as the signal of the polaron) of 70–80 mV. Therefore, the formation of the polaron is potential-delayed. It is the rare case for PANI that the polaron is not the primarily formed charged state at the polymer chain, but at first, the polaron pair is formed by a two electron transfer which can dissociate into two polarons at higher potentials. The absorption peaks of the polaron might be hidden by the absorption of the polaron pairs, but the existence of the polaron can be clearly proved by the simultaneous ESR measurements. Due to the shift of the maximum of the polaron formation from the peak current, it was concluded that the polaron can be formed both by oxidation of the neutral polymer at higher potentials and by dissociation of the polaron pair at the lower potential range.

A combined spectroelectrochemical study for poly(3-methylthiophene), poly (3-hexylthiophene) and their copolymer by ESR/UV-vis-NIR as well as FTIR spectroscopy on the influence of the copolymer composition on the stabilization of charges upon electrochemical p-doping was presented. The simultaneous use of both spectroscopies enabled the differentiation of polarons (paramagnetic) and polaron pairs (diamagnetic) in a conducting polymer [109].

The in situ ESR and UV-vis-NIR spectroelectrochemistry at higher doping levels of the polymeric materials proved bipolarons and polaron pairs as stable charged states in poly(3-hexylthiophene) as well as the copolymer copMeHeTh. During the p-doping of poly(3-methylthiophene) bipolarons were the dominating species at higher doping levels.

A combined in situ NMR and in situ ESR spectroelectrochemical study of a reaction mechanism was presented, detecting and describing the whole number of paramagnetic and diamagnetic intermediates and final products in an electrode reaction. The results of both in situ spectroelectrochemical methods at the same redox system were used to get the complete reaction mechanism [110].

Relation of conducting state to paramagnetic species has been studied by combining ESR with ac impedance. For polypyrrole, both in the oxidized conducting and neutral insulating forms exhibited a strong ESR signal [111].

Role of the formation of optically different charge carriers in the development of the conducting state have been studied using the hyphenation of a UV-VIS-NIR spectroelectrochemistry with ac conductance [112].

Plot of the conductance change versus the spectrovoltammetric data gave an answer to what an extent the development of the conducting state is connected to the formation of the given species. For poly(3-alkylthiophenes) deposited in Et_4PF_6/AN solution, the conductance developed with the spectral evolution at 1050 nm where dications/bipolarons are absorbing.

Oppositely in PEDOT (prepared in SDS/H_2O solution), this correlation has been observed for the conductance increase with the absorbance change at 770 nm, which was assigned to the excitation of monocation/polaron species.

After a detailed analysis of several systems, it has been evidenced that the charge compensating ions play a fundamental role, and fast redox switching is expected only for polymers doped by immobile anions [113].

In the case of entrapped anions, the beginning of the redox transformation generates conductance in correlation with the formation of the first independent charge carriers, called polarons or monocations. This behavior is in connection with the fact that the charge-transfer causes the migration of the cations out of the film, leaving back the anions in their proper position they occupied in the conducting state. Oppositely in anion exchanging films, the charge-transfer forces the flux of anions in from the side of the film/solution interface, and the macroscopic conductance requires the proper berth of the anions. This stage can be achieved through a properly dense distribution of the charge carriers. Thus, the onset of the conductance is delayed, and it is a result of the interaction of charge carriers in a later stage of the redox transformation, after their significant charge density has been achieved. In this case, the development of the conducting state is connected to the formation of the bipolaronic/dicationic species.

Photoelectrochemical polymer deposition under low light intensity has been studied. The results showed that this method led to a particularly low fraction of neutral PEDOT and a high fraction of bipolarons as measured in the UV-vis spectra [114].

All these information on charge carriers generated during the oxidative transformation can be summarized in a complex general scheme

Formation of (positive) polarons/monocations

$$P^0 = P^+ + e^-$$

and bipolarons/dications [1–5, 115]:

$$P^+ = B^{2+} + e^-$$

or

$$P^0 = B^{2+} + 2\,e^-$$

The interaction between polarons, leading to polaron pairs [99, 108]

$$P^+ + P^+ \rightleftarrows PP^{++} \quad \text{along the chains}$$

or to the formation of π-dimers [42]

$$P^+ + P^+ = B^{2+} \quad \text{between chains}$$

or to the disproportionation of two polarons [104]

$$P^+ + P^+ = B^{2+} + P^0$$

These processes are accompanied by ionic and solvent movements when the film contains solvent molecules as in the cases of polythiophenes. The removal of the solvent is a slow process compared to the charge transfer, and endows this section of the process with a pseudocapacitive pattern:

$$n \left[P_{sol}^+ A^- \right] \rightarrow n/2 \left[P_2^{2+} \, 2A^- \right] + n \text{ solvent}$$

which non-electrochemical process may lead to the quasi-metallic state [62].

It was assumed that this last, non-electrochemical process results in a phase transition [44], which manifests itself in the hysteresis, interpreted by the scheme-of-cube model [63].

Accordingly during the reverse process—returning from the p-doped state—the path can be other than just the opposite of the one during the oxidation [70].

In the case of the n-type transformation, chemical reversibility of the redox process and stability of the film cannot be preserved but at moderately cathodic potentials. In this restricted potential range the process leads only to negative polarons [72].

However in some cases, the attained potential region could be enough negative to result in further transformation [71], and the reoxidation from the well-reduced state took place in form of two, also spectrally distinguishable steps, demonstrating the hysteresis pattern.

It is a rational conclusion that p-type and n-type redox transformations of the given polymer follow the same mechanism although they are symmetrical for the direction of the electron transfer.

4.12 Overoxidation

Spectroelectrochemical, electrochemical, and chronovoltabsorptometric properties of conducting polymers show reversible behavior for films in a restricted potential window, but with extention of the potential range degradation may occur which is irreversible [45, 116].

Direct electropolymerization of a family of aromatic amines yielded novel, stable conducting polymers with good conductivity and solubility in organic media. Poly(N,N'-diphenyl benzidine) showed reversible behavior over several thousand cycles for pulses as short as 0.1 s, but poly(benzidine) degraded on extended cycling.

The charge-storage capacity and electrical conductivity of polypyrrole were followed through regimes of chemically reversible and irreversible electroactivity [117].

Overoxidation of polypyrrole occurs at potentials in excess of 0.7 V versus a saturated calomel electrode (SCE), as demonstrated by cyclic voltammetry of thin films. Material loss from polymer films as they are overoxidized was determined by in situ quartz microbalance experiments. The potential window for reversible electrochemistry in polypyrrole was compared to that for other conducting polymers. Reflectance FTIR of thick films revealed that hydroxyl groups, followed by carbonyls, resulted from overoxidation.

The newly developed in situ dc resistance measurement technique—called the contact electric resistance (CER) technique—as well as in situ external reflectance FTIR spectroscopy were applied for studying the conductivity and redox behavior of poly(3-methylthiophene) (PMeT) in acetonitrile and aqueous solutions [118].

The maximum conductivity as well as the effect of the thickness of the film on the evolution of the conductivity was observed to depend on whether the PMeT films are electrochemically synthesized and studied in the presence of Bu_4NClO_4 or Bu_4NPF_6 in acetonitrile. The deactivation of PMeT began in acetonitrile solutions at about 1.4 V. In accordance with this, the in situ IR spectra showed that the number of bipolarons started to decrease. The aqueous solutions were observed to decrease the maximum conductivity by 2–3 decades depending on the anion and to cause a much slower resistance response to an applied potential when compared with the results in acetonitrile. The FTIR measurements proved that in aqueous solution (bi)polarons similar to those observed in acetonitrile are formed in the film during anodic oxidation.

Although evidence has been found that the cycling of PPy films up to 4.6–4.8 V is reversible for the first few cycles, irreversible oxidation took place later because the evolving dications located on the polymeric chains might be subjected to nucleophilic attacks from the residual molecules of H_2O or HCO_3^- ions, in good agreement with the XPS data obtained [102].

The behavior of poly(3,4-butylenedioxythiophene) (PBuDOT), a relative of poly (3,4-ethylenedioxythipohnene) PEDOT within the poly(3,4-alkylenedioxythio-phene) family, has been studied at potentials above its electrochemical stability threshold using in situ ESR spectroelectrochemistry [119].

The aim was to investigate the effect of electrochemical overoxidation on the charge-carrying species, namely polarons by determining the potential dependencies of spectroscopic parameters of the ESR spectra of the polymer over a selected potential range. Specific features of the trends of these dependencies allowed also the evaluation of presence of the second type of charge-carrying species—dia-magnetic bipolarons—and the effects of their interactions with polarons at different potentials. Around 1.5 V, where the boundary of electrochemical stability of the polymer lies, sharp drop of the concentration of paramagnetic centers has been observed together with a transitory narrowing of the ESR line. These changes were found to be irreversible as evidenced by the course of subsequent reduction half-cycle. It was concluded that the overoxidation process leads to a degradation of the polymer most probably due to a decrease of the conjugation length of the main chain π-bond through cross-linking. While the electrochemical results pointed to a non-complete degradation of the polymer, the specific parameters of the ESR line in the reduction half-cycle indicated that the remaining spins are confined to isolated segments of a partially degraded polymer where their behavior resembles oligomer-like radicals.

It is important to note that overoxidized polymers and their composites can be used as combined electrodes for electroanalytical purposes [120, 121].

4.13 Structural Questions

Reversible thermochromism in thin solid films of poly(3-hexylthiophene) has been studied using ultraviolet and X-ray photoelectron spectroscopies [122].

The analysis of the spectra indicated that at elevated temperatures thermally induced electronic localization occurred as a consequence of the induced confor-mational disorder.

The structure of emeraldine salt and emeraldine bases with different molar weight and their behavior in electrochemical doping was studied by different spectroscopic and spectroelectrochemical techniques [123].

By Fourier transform infrared (FTIR) spectroscopy, the branching of the polymer chain at tri- and tetra-substituted benzene rings as well as the presence of small amounts of phenazine units have been shown. The branching of the polymer chains increased with the increase in the molar weight of emeraldines.

The combination of reflectance UV-Vis spectroelectrochemistry with electrochemical quartz crystal microbalance (EQCM) revealed that the polyaniline-like chains in poly(o-phenylene diamine) could be completely converted via intramolecular cyclization into the ladder structure with phenazine units [124].

4.14 Miscellaneous Issues

On the water uptake/extruding and diffusion in polymer films

The problem was studied by combined EIS and FTIR–ATR study in various polymer films on semiconducting electrodes [125].

Based on studies of pyrrole polymerization in an aqueous medium by in situ ellipsometry and FTIR spectroscopy, the expulsion of solvated protons were concluded [45] while the solvent removal during electrochemical polymerization of pyrrole [26] and the oxidation of thiophene type CP films was concluded from EQCM measurements [64].

Interaction with the substrate during the initial stages of the electrodeposition

Results of a scanning tunneling microscopy (STM) and photogalvanic study of the growth of poly(3-methylthiophene) on indium tin oxide (ITO) electrodes was presented. ITO itself exhibited irregular surface structure consisting of grains, separated by deep recesses. Polymer growth began as oligomer islets on top of the grains and proceeded to form a thin film. Several polymeric features, like fibrils and large featureless bundles, could be detected during later stages of polymerization. Finally macroscopic growth centers appeared on the surface and their evolution determines the structure of a thick film [126].

Electrochemical growth of conducting polymer polythiophene was investigated by X-ray photoemission spectroscopy (XPS) [127].

The initial stage of polymer growth was investigated by analyzing the core-level energies and spectral profiles of the atomic components. The experimental result reflected the interface linkage between the organic polymer chain and the Si oxidized layer such as S–O–Si.

A nanoscale-linked-crater structure was fabricated on an Al surface by chemical and electrochemical combination processes, and the thin film of conducting polymer polythiophene was grown on the surface by an electrochemical method, and studied by dynamic force microscopy and X-ray photoemission spectroscopy [128].

Interaction with the solvent during the electrodeposition

The influence of deposition conditions on solvation of reactive conducting polymers has been studied by in situ neutron reflectivity (NR) and reflection–absorption infrared spectroscopy (RAIRS) [129].

The role of the solvent in controlling the rate of reaction was explored. Deposition rate yielded films containing different solvent content: rapid deposition gave films with a more open structure leading to a higher solvent level.

Reflectance imaging has been successfully applied for in situ monitoring of the drying process of film formation for organic photovoltaics over large areas [130].

The drying wet film was illuminated with a narrow bandwidth LED with the specularly reflected light recorded by a video camera as the film dried and formed the active layer of the OPV cell.

Electromagnetic shielding properties

The electromagnetic shielding behavior of conducting polymers received much attention for practical reasons. A recent paper illuminated well the capabilities of conducting polymers with excellent electromagnetic interference (EMI) properties [131].

Further remarkable papers reporting on the electromagnetic shielding effect of conducting polymers are also available [132–145].

References

1. Garnier F, Tourillon T, Gazard M, Dubois JC (1983) Organic conducting polymers derived from substituted thiophenes as electrochromic material. J Electroanal Chem 148:299–303. doi:10.1016/S0022-0728(83)80406-9
2. Genies EM, Bidan G, Diaz AF (1983) Spectroelectrochemical study of polypyrrole films. J Electroanal Chem 149:101–113. doi:10.1016/S0022-0728(83)80561-0
3. Kobayashi T, Yoneyama H, Tamura H (1984) Polyaniline film-coated electrodes as electrochromic display devices. J Electroanal Chem 161:419–423. doi:10.1016/S0022-0728 (84)80201-6
4. Kuwabata S, Yoneyama H, Tamura H (1984) Redox behavior and electrochromic properties of polypyrrole films in aqueous-solutions. Bull Chem Soc Jpn 57:2247–2253. doi:10.1246/ bcsj.57.2247
5. Inganas O, Lundstrom I (1984) A photoelectrochromic memory and display device based on conducting polymers. J Electrochem Soc 131:1129–1132. http://apps.webofknowledge.com/ full_record.do?product=WOS&search_mode=GeneralSearch&qid=1&SID=Q1qEYwxHJ 133rClpT9s&page=1&doc=1. Accessed 5 Dec 2016
6. Christensen PA, Hamnett A, Hillman AR (1988) An in situ infrared study of polythiophene growth. J Electroanal Chem 242:47–62. doi:10.1016/0022-0728(88)80238-9
7. Bacskai J, Inzelt G, Bartl A, Dunsch L, Paasch G (1994) In situ electrochemical ESR investigations of the growth of one- and two-dimensional polypyrrole films. Synth Met 67:227–230. doi:10.1016/0379-6779(94)90046-9
8. Roncali J, Guy A, Lemaire M, Garreau R, Hoa HA (1991) Tetrathienylsilane as a precursor of highly conducting electrogenerated polythiophene. J Electroanal Chem 312:277–283. doi:10.1016/0022-0728(91)85159-M

9. Visy C, Lukkari J, Kankare J (1996) Electrochemical polymerization of the tetrathienyl derivatives of the carbon group elements. J Electroanal Chem 401:119–125. doi:10.1016/0022-0728(95)04261-X

10. Zimmermann A, Künzelmann U, Dunsch L (1998) Initial states in the electropolymerization of aniline and p-aminodiphenylamine as studied by in situ FT-TR and UV-Vis spectroelectrochemistry. Synth Met 93:17–25. doi:10.1016/S0379-6779(98)80127-6

11. Skompska M (2000) Quartz crystal microbalance study of electrochemical deposition of poly(3-dodecylthiophene) films on Au electrodes. Electrochim Acta 45:3841–3850. doi:10.1016/S0013-4686(00)00457-6

12. Wang LQ, Kowalik J, Mizaikoff B, Kranz C (2010) Combining scanning electrochemical microscopy with infrared attenuated total reflection spectroscopy for in situ studies of electrochemically induced processes. Anal Chem 82:3139–3145. doi:10.1021/ac9027802

13. Liu M, Ye M, Yang Q, Zhang Y, Xie Q, Yao S (2006) A new method for characterizing the growth and properties of polyaniline and poly(aniline-co-o-aminophenol) films with the combination of EQCM and in situ FTIR spectroelectrochemistry. Electrochim Acta 52: 342–352. doi:10.1016/j.electacta.2006.05.013

14. Rishpon J, Gottesfeld S (1991) Investigation of polypyrrole glucose-oxidase electrodes by ellipsometric, microgravimetric and electrochemical measurements. Biosens Bioelectron 6:143–149. doi:10.1016/0956-5663(91)87038-D

15. Shim HS, Yeo IH, Park SM (2002) Simultaneous multimode experiments for studies of electrochemical reaction mechanisms: demonstration of concept. Anal Chem 74:3540–3546. doi:10.1021/ac020121c

16. Jiang L, Xie QJ, Yang L, Yang XY, Yao SZ (2004) Simultaneous EQCM and diffuse reflectance UV-visible spectroelectrochemical measurements: poly(aniline-co-o-anthranilic acid) growth and property characterization. J Colloid Interface Sci 274:150–158. doi:10.1016/j.jcis.2004.01.044

17. Visy C, Lukkari J, Kankare J (1997) Change from a bulk to a surface coupling mechanism in the electrochemical polymerization of thiophene. Synth Met 87:81–87. doi:10.1016/S0379-6779(97)80101-4

18. Hamnett A, Hillman AR (1987) Ellipsometric study of the growth of thin organic polymer-films. Berichte der Bunsengesellschaft/Phys Chem Chem Phys 91:329–336. http://apps.webofknowledge.com/full_record.do?product=WOS&search_mode=GeneralSearch&qid=51&SID=Q1qEYwxHJ133rClpT9s&page=1&doc=1. Accessed 5 Dec 2016

19. Kim YT, Collins RW, Vedam K, Allara DL (1991) Real-time spectroscopic ellipsometry—in situ characterization of pyrrole electropolymerization. J Electrochem Soc 138:3266–3275. doi:10.1149/1.2085401

20. Li FB, Albery WJ (1992) A novel mechanism of electrochemical deposition of conducting polymers—2-dimensional layer-by-layer nucleation and growth observed for poly(thiophene-3-acetic acid). Electrochim Acta 37:393–401. doi:10.1016/0013-4686(92)87027-W

21. Raymond DE, Harrison DJ (1993) Observation of soluble pyrrole oligomers and the role of protons in the formation of polypyrrole and polybipyrrole. J Electroanal Chem 355:115–131. doi:10.1016/0022-0728(93)80357-N

22. Raymond DE, Harrison DJ (1993) Observation of pyrrole radical cations as transient intermediates during the anodic formation of conducting polypyrrole films. J Electroanal Chem 361:65–76. doi:10.1016/0022-0728(93)87039-X

23. Visy C, Lukkari J, Kankare J (1994) Study of the role of the deprotonation step in the electrochemical polymerization of thiophene-type monomers. Synth Met 66:61–65. doi:10.1016/0379-6779(94)90162-7

24. Lukkari J, Kankare J, Visy C (1992) Cyclic spectrovoltammetry: a new method to study the redox processes in conductive polymers. Synth Met 48:181–192. doi:10.1016/0379-6779(92)90060-V

25. Kriván E, Visy C (2001) New phenomena observed during the electrochemical reduction of conducting polypyrrole films. J Solid State Electrochem 5:507–511. http://apps.web ofknowledge.com/full_record.do?product=WOS&search_mode=GeneralSearch&qid=58& SID=Q1qEYwxHJ133rClpT9s&page=1&doc=1. Accessed 5 Dec 2016

26. Kriván E, Visy C, Kankare J (2003) Deprotonation and dehydration of pristine PPy/DS films during the open-circuit relaxation: an ignored factor in determining the properties of conducting polymers. J Phys Chem B 107:1302–1308. doi:10.1021/jp0214615

27. Tourillon G, Garnier F (1983) Effect of dopant on the physicochemical and electrical-properties of organic conducting polymers. J Phys Chem 87:2289–2292. doi:10.1021/ j100236a010

28. Visy C, Lukkari J, Pajunen T, Kankare J (1990) Spectroelectrochemical study of the anion effect on the transient redox behavior of poly(n-methylpyrrole) in anhydrous acetonitrile. Synth Met 39:61–67. doi:10.1016/0379-6779(90)90198-T

29. Visy C, Lukkari J, Pajunen T, Kankare J (1989) Effect of anions on the transient redox behavior of polypyrrole in anhydrous acetonitrile. Synth Met 33:289–299. doi:10.1016/ 0379-6779(89)90475-X

30. Tóth PS, Janáky C, Berkesi O, Tamm T, Visy C (2012) On the unexpected cation exchange behavior, caused by covalent bond formation between PEDOT and Cl⁻ ions, extending the conception for polymer—dopant interactions. J Phys Chem B 116:5491–5500. doi:10.1021/ jp2107268

31. Tóth PS, Endrődi B, Janaky C, Visy C (2015) Development of polymer–dopant interactions during electropolymerization, a key factor in determining the redox behaviour of conducting polymers. J Solid State Electrochem 19:2891–2896. doi:10.1007/s10008-015-2791-1

32. Visy C, Bencsik G, Nemeth Z, Vertes A (2008) Synthesis and characterization of chemically and electrochemically prepared conducting polymer/iron oxalate composites. Electrochim Acta 53:3942–3947. doi:10.1016/j.electacta.2007.07.060

33. Cs Janaky, Visy C, Berkesi O, Tombacz E (2009) Conducting polymer based electrode with magnetic behavior: electrochemical synthesis of poly(3-thiophene-aceticacid)/magnetite nanocomposite thin layers. J Phys Chem C 113:1352–1358. doi:10.1021/jp809345b

34. Endrődi B, Bíró A, Janaky C, Toth IY, Visy C (2013) Layer by layer growth of electroactive conducting polymer/magnetite hybrid assemblies. Synth Met 171:62–68. doi:10.1016/j. synthmet.2013.03.016

35. Cs Janáky, Endrődi B, Kovacs K, Timko M, Sapi A, Visy C (2010) Chemical synthesis and characterization of poly(3-thiophene-acetic acid)/magnetite nanocomposites with tunable magnetic behaviour. Synth Met 160:65–71. doi:10.1016/j.synthmet.2009.09.034

36. Janáky C, Visy C (2013) Conducting polymer-based hybrid assemblies for electrochemical sensing: a materials science perspective. Anal Bioanal Chem 405:3489–3511. doi:10.1007/ s00216-013-6702-y

37. Janáky C, Endrődi B, Hajdu A, Visy C (2010) Synthesis and characterization of polypyrrole–magnetite–vitamin B12 hybrid composite electrodes. J Solid State Electrochem 14:339–346. doi:10.1007/s10008-009-0827-0

38. Krukiewicz K, Bednarczyk-Cwynar B, Turczyn R, Zak JK (2016) EQCM verification of the concept of drug immobilization and release from conducting polymer matrix. Electrochim Acta 212:694–700. doi:10.1016/j.electacta.2016.07.055

39. Dahlin AB, Dielacher B, Rajendran P, Sugihara K, Sannomiya T, Zenobi-Wong M, Voros J (2012) Electrochemical plasmonic sensors. Anal Bioanal Chem 402:1773–1784. doi:10. 1007/s00216-011-5404-6

40. Patil AO, Heeger AJ, Wudl F (1988) Optical properties of conducting polymers. Chem Rev 88:183–200. doi:10.1021/cr00083a009

41. Fichou D, Horowitz G, Garnier F (1990) Polaron and bipolaron formation on isolated-model thiophene oligomers in solution. Synth Met 39:125–131. doi:10.1016/0379-6779(90)90207-2

42. Hill MG, Penneau Zinger B, Mann KR, Miller LL (1992) Oligothiophene cation radicals— pi-dimers as alternatives to bipolarons in oxidized polythiophenes. Chem Mater 4: 1106–1113. doi:10.1021/cm00023a032

43. Heinze J, Tschuncky P, Smie A (1998) The oligomeric approach—the electrochemistry of conducting polymers in the light of recent research. J Solid State Electrochem 2:102–109. doi:10.1007/s100080050073

44. Lapkowski M, Zagorska M, Kulszewiczbajer I, Koziel K, Pron A (1991) Spectroelectrochemical properties of poly(4,4'-dialkyl-2,2'-bithiophenes) and poly (3-alkylthiophenes)—a comparative-study. J Electroanal Chem 310:57–70. http://apps.web ofknowledge.com/full_record.do?product=WOS&search_mode=GeneralSearch&qid=61& SID=Q1qEYwxHJ133rClpT9s&page=1&doc=3. Accessed 5 Dec 2016

45. Christensen PA, Hamnett A (1991) In situ spectroscopic investigations of the growth, electrochemical cycling and overoxidation of polypyrrole in aqueous-solution. Electrochim Acta 36:1263–1286. doi:10.1016/0013-4686(91)80005-S

46. Kellenberger A, Dmitrieva E, Dunsch L (2012) Structure dependence of charged states in "linear" polyaniline as studied by in situ ATR-FTIR spectroelectrochemistry. J Phys Chem B 116:4377–4385. doi:10.1021/jp211595n

47. Zhang WB, Dong SJ (1993) Effects of dopant and solvent on the properties of polypyrrole—investigations with cyclic voltammetry and electrochemically in situ conductivity. Electrochim Acta 38:441–445. doi:10.1016/0013-4686(93)85163-S

48. Salavagione HJ, Arias-Pardilla J, Perez JM, Vazquez JL, Morallon E, Miras MC, Barbero C (2005) Study of redox mechanism of poly(o-aminophenol) using in situ techniques: evidence of two redox processes. J Electroanal Chem 576:139–145. doi:10.1016/j.jelechem. 2004.10.013

49. Kocharova N, Lukkari J, Viinikanoja A, Aaritalo T, Kankare J (2002) Doping-induced structural changes of conducting polyalkoxythiophene on the chemically modified gold surface: an in situ surface enhanced resonance Raman spectroscopic study. J Phys Chem B 106:10973–10981. doi:10.1021/jp026259g

50. Vieil E, Meerholz K, Matencio T, Heinze J (1994) Mass transfer and convolution. 2. In situ optical beam deflection study of ionic exchanges between polyphenylene films and a 1:1 electrolyte. J Electroanal Chem 368:183–191. doi:10.1016/0022-0728(93)03110-B

51. Vorotyntsev MA, Lopez C, Vieil E (1994) On the interpretation of optical beam deflection data at excess of a background electrolyte. J Electroanal Chem 368:155–163. doi:10.1016/ 0022-0728(93)03095-7

52. Lopez C, Viegas MFM, Bidan G, Vieil E (1994) Comparison of ion exchange properties of polypyrrole with and without immobilized dopants by optical beam deflection. Synth Met 63:73–78. doi:10.1016/0379-6779(94)90252-6

53. Zotti G, Schiavon G (1991) Spin and spinless conductivity in polypyrrole. Evidence for mixed-valence conduction. Synth Met 41:445–448. doi:10.1016/0379-6779(91)91104-I

54. Tanguy J, Mermilliod N, Hoclet M (1987) The capacitive charge and noncapacitive charge in conducting polymer electrodes. J Electrochem Soc 134:795–802. doi:10.1149/1.2100575

55. Tanguy J, Baudoin JL, Chao F, Costa M (1992) Study of the redox mechanism of poly-3-methylthiophene by impedance spectroscopy. Electrochim Acta 37:1417–1428. doi:10.1016/0013-4686(92)87016-S

56. Visy C, Lakatos M, Szucs A, Novak M (1997) Separation of faradaic and capacitive current regions in the redox transformation of poly(3-methylthiophene) with the exclusion of overoxidation processes. Electrochim Acta 42:651–657. doi:10.1016/S0013-4686(96) 00210-1

57. Kalaji M, Peter LM (1991) Optical and electrical ac response of polyaniline films. J Chem Soc, Faraday Trans 87:853–860. doi:10.1039/ft9918700853

58. Higgins SJ, Hamnett A (1991) In situ ellipsometric study of the growth and electrochemical cycling of polypyrrole films on platinum. Electrochim Acta 36:2123–2134. doi:10.1016/ 0013-4686(91)85220-2

59. Suarez MF, Compton RG (1999) In situ atomic force microscopy study of polypyrrole synthesis and the volume changes induced by oxidation and reduction of the polymer. J Electroanal Chem 462:211–221. doi:10.1016/S0022-0728(98)00414-8

60. Hillman AR, Swann MJ, Bruckenstein S (1990) Ion and solvent transfer accompanying polybithiophene doping and undoping. J Electroanal Chem 291:147–162. doi:10.1016/0022-0728(90)87183-K

61. Otero TF, Rodriguez J, Angulo E, Santamaria C (1993) Artificial muscles from bilayer structures. Synth Met 57:3713–3717. doi:10.1016/0379-6779(93)90502-N

62. Visy C, Kankare J (2000) Direct in situ conductance evidence for non-faradaic electrical processes in poly(3-methylthiophene). Electrochim Acta 45:1811–1820. doi:10.1016/S0013-4686(99)00404-1

63. Hillman AR, Bruckenstein S (1993) Role of film history and observational timescale in redox switching kinetics of electroactive films. 1. A new model for permselective films with polymer relaxation processes. J Chem Soc, Faraday Trans 89:339–348. doi:10.1039/ft9938900339

64. Visy C, Kankare J, Krivan E (2000) EQCM and in situ conductance studies on the polymerization and redox features of thiophene co-polymers. Electrochim Acta 45:3851–3864. doi:10.1016/S0013-4686(00)00456-4

65. Krivàn E, Visy C, Kankare J (2005) Key role of the desolvation in the achievement of the quasi-metallic state of electronically conducting polymers. Electrochim Acta 50:1247–1254. doi:10.1016/j.electacta.2004.07.050

66. Mastragostino M, Soddu L (1990) Electrochemical characterization of n-doped polyheterocyclic conducting polymers. 1. Polybithiophene. Electrochim Acta 35:463–466. doi:10.1016/0013-4686(90)87029-2

67. Visy C, Lukkari J, Pajunen T, Kosonen J, Kankare J (1989) Spectroscopic evidence for the existence of long-lived intermediates during the electrochemical transformation of poly-3-methylthiophene. J Electroanal Chem 262:297–301. doi:10.1016/0022-0728(89) 80031-2

68. Hillman AR, Loveday DC, Moffatt DE, Maher J (1992) In situ electron-paramagnetic resonance-spectra of n-doped and p-doped poly(benzo[c]thiophene) films. J Chem Soc, Faraday Trans 88:3383–3384. doi:10.1039/ft9928803383

69. Neugebauer H, Cravino A, Luzzati S, Catellani M, Petr A, Dunsch L, Sariciftci NS (2003) Spectral signatures of positive and negative charged states in doped and photoexcited low band-gap poly-dithienothiophenes. Synth Met 139:747–750. doi:10.1016/S0379-6779(03) 00292-3

70. Visy C, Lukkari J, Kankare J (1993) Scheme for the anodic and cathodic transformations in polythiophenes. Macromolecules 26:3295–3298. doi:10.1021/ma00065a008

71. Visy C, Lukkari J, Kankare J (1994) Electrochemically polymerized terthiophene derivatives carrying aromatic substituents. Macromolecules 27:3322–3329. doi:10.1021/ma00090a028

72. Ahonen HJ, Lukkari J, Kankare J (2000) N- and p-doped poly(3,4-ethylenedioxythiophene): two electronically conducting states of the polymer. Macromolecules 33:6787–6793. doi:10.1021/ma0004312

73. Roncali J (2007) Molecular engineering of the band gap of p-conjugated systems: facing technological applications. Macromol Rapid Commun 28:1761–1775. doi:10.1002/marc. 200700345

74. Beyer R, Kalaji M, Kingscote-Burton G, Murphy PJ, Pereira VMSC, Taylor DM, Williams GO (1998) Spectroelectrochemical and electrical characterization of low bandgap polymers. Synth Met 92:25–31. doi:10.1016/S0379-6779(98)80018-0

75. Cravino A, Neugebauer Luzzati S, Catellani M, Petr A, Dunsch L, Sariciftci NS (2002) Positive and negative charge carriers in doped or photoexcited polydithienothiophenes: a comparative study using Raman, infrared, and electron spin resonance spectroscopy. J Phys Chem B 106:3583–3591. doi:10.1021/jp013351o

76. Damlin P, Kvarnström C, Petr A, Ek P, Dunsch L, Ivaska A (2002) In situ resonant Raman and ESR spectroelectrochemical study of electrochemically synthesized poly (p-phenylenevinylene). J Solid State Electrochem 6:291–301. doi:10.1007/s100080100240

77. Sezer E, Skompska M, Heinze J (2008) Voltammetric, EQCM, and in situ conductance studies of p- and n-dopable polymers based on ethylenedioxythiophene and bithiazole. Electrochim Acta 53:4958–4968. doi:10.1016/j.electacta.2008.02.027

78. Pappenfus TM, Schneiderman DK, Casado J, Navarrete JTL, Delgado MCR, Zotti G, Vercelli B, Lovander MD, Hinkle LM, Bohnsack JN, Mann KR (2011) Oligothiophene tetracyanobutadienes: alternative donor-acceptor architectures for molecular and polymeric materials. Chem Mater 23:823–831. doi:10.1021/cm102128g

79. Abbotto A, Calderon EH, Manfredi N, Mari CM, Marinzi C, Ruffo R (2011) Vinylene-linked pyridine-pyrrole donor-acceptor conjugated polymers. Synth Met 161:763–769. doi:10.1016/j.synthmet.2011.01.027

80. Li W, Lee T, Oh SJ, Kagan CR (2011) Diketopyrrolopyrrole-based pi-bridged donor-acceptor polymer for photovoltaic applications. ACS Appl Mater Interfaces 3:3874–3883. doi:10.1021/am200720e

81. Meana-Esteban B, Balan A, Baran D, Neugebauer H, Toppare L, Sariciftci NS (2011) In situ spectroelectrochemical study of positively and negatively charged states in a donor/acceptor EDOT/benzotriazole-based polymer. Macromol Chem Phys 212:2459–2466. doi:10.1002/macp.201100322

82. Feng X, Liu L, Honsho Y, Saeki A, Seki S, Irle S, Dong YP, Nagai A, Jiang DL (2012) High-rate charge-carrier transport in porphyrin covalent organic frameworks: switching from hole to electron to ambipolar conduction. Angewandte Chemie-Interned 51:2618–2622. doi:10.1002/anie.201106203

83. Khatib O, Yuen JD, Wilson J, Kumar R, Di Ventra M, Heeger AJ, Basov DN (2012) Infrared spectroscopy of narrow gap donor-acceptor polymer-based ambipolar transistors. Phys Rev B 86:195109. doi:10.1103/PhysRevB.86.195109

84. Lav TX, Tran-Van F, Aubert PH, Chevrot C (2012) Elaboration and characterization of donor-acceptor polymer through electropolymerization of fullerene substituted N-alkylcarbazole. Synth Met 162:1923–1929. doi:10.1016/j.synthmet.2012.09.003

85. Camurlu P, Karagoren N (2013) Both p and n-dopable, multichromic, napthalineimide clicked poly(2,5-dithienylpyrrole) derivatives. J Electrochem Soc 160:H560–H567. doi:10.1149/2.043309jes

86. Baeg KJ, Caironi M, Noh YY (2013) Toward printed integrated circuits based on unipolar or ambipolar polymer semiconductors. Adv Mater 25:4210–4244. doi:10.1002/adma.201205361

87. Meana-Esteban B, Petr A, Kvarnstrom C, Ivaska A, Dunsch L (2014) Poly (2-methoxynaphthalene): a spectroelectrochemical study on a fused ring conducting polymer. Electrochim Acta 115:10–15. doi:10.1016/j.electacta.2013.10.090

88. Ledwon P, Thomson N, Angioni E, Findlay NJ, Skabara PJ, Domagala W (2015) The role of structural and electronic factors in shaping the ambipolar properties of donor-acceptor polymers of thiophene and benzothiadiazole. RSC Adv 5:77303–77315. doi:10.1039/c5ra06993a

89. Laba K, Data P, Zassowski P, Pander P, Lapkowski M, Pluta K, Monkman AP (2015) Diquinoline derivatives as materials for potential optoelectronic applications. J Phys Chem C 119:13129–13137. doi:10.1021/jp512941z

90. Kaufman JH, Colaneri N, Scott JC, Kanazawa KK, Street GB (1985) Evolution of polaron states into bipolarons in polypyrrole. Mol Cryst Liq Cryst 118:171–177. doi:10.1080/00268948508076206

91. Leclerc M (1990) Characterization of a bipolaronic form in poly(2-methylaniline). J Electroanal Chem 296:93–100. doi:10.1016/0022-0728(90)87235-C

92. Lefrant S, Buisson JP, Eckhardt H (1990) Raman spectra of conducting polymers with aromatic rings. Synth Met 37:91–98. doi:10.1016/0379-6779(90)90131-4

93. Amemiya T, Hashimoto K, Fujishima A, Itoh K (1991) Analyses of spectroelectrochemical behavior of polypyrrole films using the Nernst-equation—monomer unit model and polaron/bipolaron model. J Electrochem Soc 138:2845–2859. doi:10.1149/1.2085327

94. Visy C, Lukkari J, Kankare J (1991) A thermodynamic approach to the interpretation of anodic and cathodic doping of poly(3-methylthiophene). J Electroanal Chem 319:85–100. doi:10.1016/0022-0728(91)87069-G

95. Kankare J, Lukkari J, Pajunen T, Ahonen J, Visy C (1990) Evolutionary spectral factor analysis of doping-undoping processes of thin conductive polymer films. J Electroanal Chem 294:59–72. doi:10.1016/0022-0728(90)87135-7

96. Son Y, Rajeshwar K (1992) Potential-modulated ultraviolet visible and raman-spectra of polypyrrole thin-films in aqueous-electrolytes—combination with voltammetric scanning and the influence of dioxygen on the stability of radical cations and dications of the conducting polymer. J Chem Soc, Faraday Trans 88:605–610. doi:10.1039/ft9928800605

97. Trznadel M, Zagorska M, Lapkowski M, Louarn G, Lefrant S, Pron A (1996) UV-VIS-NIR and Raman spectroelectrochemistry of regioregular poly(3-octylthiophene): comparison with its non-regioregular analogue. J Chem Soc, Faraday Trans 92:1387–1393. doi:10.1039/ft9969201387

98. Petr A, Dunsch L (1996) Direct evidence of indamine cation radicals in the anodic oxidation of aniline by in situ ESR spectroscopy. J Electroanal Chem 419:55–59. doi:10.1016/S0022-0728(96)04861-9

99. Rapta P, Neudeck A, Petr A, Dunsch L (1998) In situ EPR/UV-VIS spectroelectrochemistry of polypyrrole redox cycling. J Chem Soc, Faraday Trans 94:3625–3630. doi:10.1039/a806423g

100. Neudeck A, Petr A, Dunsch L (1999) Separation of the ultraviolet-visible spectra of the redox states of conducting polymers by simultaneous use of electron-spin resonance and ultraviolet-visible spectroscopy. J Phys Chem B 103:912–919. doi:10.1021/jp983383k

101. Lapkowski M, Pron A (2000) Electrochemical oxidation of poly(3,4-ethylene-dioxythiophene)—"in situ" conductivity and spectroscopic investigations. Synth Met 110:79–83. doi:10.1016/S0379-6779(99)00271-4

102. Levi MD, Lankri E, Gofer Y, Aurbach D, Otero T (2002) The behavior of polypyrrole-coated electrodes in propylene carbonate solutions—I. Characterization of PPy films by a combination of electroanalytical tools and XPS. J Electrochem Soc 149: E204–E214. doi:10.1149/1.1475691

103. Zykwinska A, Domagala W, Czardybon A, Pilawa B, Lapkowski M (2003) In situ EPR spectroelectrochemical studies of paramagnetic centres in poly(3,4-ethylenedioxythiophene) (PEDOT) and poly(3,4-butylenedioxythiophene) (PBuDOT) films. Chem Phys 292:31–45. doi:10.1016/S0301-0104(03)00253-2

104. Rapta P, Lukkari J, Tarabek J, Salomaki M, Jussila M, Yohannes G, Riekkola ML, Kankare J, Dunsch L (2004) Ultrathin polyelectrolyte multilayers: in situ ESR/UV-Vis-NIR spectroelectrochemical study of charge carriers formed under oxidation. Phys Chem Chem Phys 6:434–441. doi:10.1039/b308891j

105. Paasch G, Scheinert S, Petr A, Dunsch L (2006) Bipolarons or polaron pairs in conducting polymers: equilibrium and kinetics. Russ J Electrochem 42:1161–1168. doi:10.1134/S1023193506110024

106. Wei D, Espindola P, Lindfors T, Kvarnstrom C, Heinze J, Ivaska A (2007) In situ conductance and in situ ATR-FTIR study of poly(N-methylaniline) in aqueous solution. J Electroanal Chem 602:203–209. doi:10.1016/j.jelechem.2006.12.017

107. Di W, Petr A, Kvarnstrom C, Dunsch L, Ivaska A (2007) Charge carrier transport and optical properties of poly[N-methyl(aniline)]. J Phys Chem C 111:16571–16576. doi:10.1021/jp074712o

108. Dmitrieva E, Harima Y, Dunsch L (2009) Influence of phenazine structure on polaron formation in polyaniline: in situ electron spin resonance-ultraviolet/visible-near-infrared spectroelectrochemical study. J Phys Chem B 113:16131–16141. doi:10.1021/jp9072944

109. Chazaro-Ruiz LF, Kellenberger A, Dunsch L (2009) In situ ESR/UV-vis-NIR and ATR-FTIR spectroelectrochemical studies on the p-doping of copolymers of 3-methyl-thiophene and 3-hexylthiophene. J Phys Chem B 113:2310–2316. doi:10.1021/jp806810r

110. Klod S, Dunsch L (2011) A combination of in situ ESR and in situ NMR spectroelectro-chemistry for mechanistic studies of electrode reactions: the case of p-benzoquinone. Magn Reson Chem 49:725–729. doi:10.1002/mrc.2819

111. Waller AM, Compton RG (1989) In-situ electrochemical ESR. Compr Chem Kinet 29: 297–352. 10.1016/S0069-8040(08)70322-4. Accessed 5 Dec 2016

112. Tóth PS, Peintler-Kriván E, Visy C (2010) Application of simultaneous monitoring of the in situ impedance and optical changes on the redox transformation of two polythiophenes: direct evidence for their non-identical conductance—charge carrier correlation. Electrochem Commun 12:958–961. doi:10.1016/j.elecom.2010.05.001

113. Tóth PS, Peintler-Kriván E, Visy C (2012) Fast redox switching into the conducting state, related to single mono-cationic/polaronic charge carriers only in cation exchanger type conducting polymers. Electrochem Commun 18:16–19. doi:10.1016/j.elecom.2012.02.005

114. Park BW, Yang L, Johansson EMJ, Vlachopoulos N, Chams A, Perruchot C, Jouini M, Boschloo G, Hagfeldt A (2013) Neutral, polaron, and bipolaron states in PEDOT prepared by photoelectrochemical polymerization and the effect on charge generation mechanism in the solid-state dye-sensitized solar cell. J Phys Chem C 117:22484–22491. doi:10.1021/jp406493v

115. Brédas JL, Themans B, Fripiat JG, Andre JM, Chance RR (1984) Highly conducting polyparaphenylene, polypyrrole and polythiophene chains—an abinitio study of the geometry and electronic-structure modifications upon doping. Phys Rev B 29:6761–6773. doi:10.1103/PhysRevB.29.6761

116. Chandrasekhar P, Gumbs RW (1991) Electrosyntheses, spectroelectrochemical, electro-chemical, and chronovoltabsorptometric properties of family of poly(aromatic-amines), novel processible conducting polymers. 1. Poly(benzidines). J Electrochem Soc 138: 1337–1346

117. Schlenoff JB, Xu H (1992) Evolution of physical and electrochemical properties of polypyrrole during extended oxidation. J Electrochem Soc 139:2397–2401. doi:10.1149/1.2221238

118. Lankinen E, Sundholm G, Talonen P, Laitinen T, Saario T (1998) Characterization of a poly (3-methylthiophene) film by an in-situ dc resistance measurement technique and in-situ FTIR spectroelectrochemistry. J Electroanal Chem 447:135–145. doi:10.1016/S0022-0728 (98)00012-6

119. Zykwinska A, Domagala W, Czardybon A, Pilawa B, Lapkowski M (2006) In-situ ESR spectroelectrochemical studies of overoxidation behaviour of poly (3,4-butylenedioxythiophene). Electrochim Acta 51:2135–2144. doi:10.1016/j.electacta.2005.03.081

120. Ge H, Zhang J, Wallace GG (1992) Use of overoxidized polypyrrole as a chromium(VI) sensor. Anal Lett 25:429–441. http://apps.webofknowledge.com/full_record.do?product= WOS&search_mode=GeneralSearch&qid=114&SID=Q1qEYwxHJ133rClpT9s&page=1& doc=1. Accessed 5 Dec 2016

121. Gao YS, Xu JK, Lu LM, Wu LP, Zhang KX, Nie T, Zhu XF, Wu Y (2014) Overoxidized polypyrrole/graphene nanocomposite with good electrochemical performance as novel electrode material for the detection of adenine and guanine. Biosens Bioelectron 62: 261–267. doi:10.1016/j.bios.2014.06.044

122. Salaneck WR, Inganas O, Themans B, Nilsson JO, Sjogren B, Osterholm JE, Bredas JL, Svensson S (1988) Thermochromism in poly(3-hexylthiophene) in the solid-state—a spectroscopic study of temperature-dependent conformational defects. J Chem Phys 89:4613–4619. doi:10.1063/1.454802

123. Dmitrieva E, Dunsch L (2011) How linear is "linear" polyaniline? J Phys Chem B 115:6401–6411. doi:10.1021/jp200599f

124. Tu XM, Xie QJ, Wang ML, Zhang YY, Yao SZ (2005) Interconversion for a polymer between its ladder and linear structure. Chin Sci Bull 50:1598–1604. doi:10.1360/982004-681

125. Vlasak R, Klueppel I, Grundmeier G (2007) Combined EIS and FTIR–ATR study of water uptake and diffusion in polymer films on semiconducting electrodes. Electrochim Acta 52:8075–8080. doi:10.1016/j.electacta.2007.07.003

126. Lukkari J, Alanko M, Heikkila L, Laiho R, Kankare J (1993) Nucleation and growth of poly (3-methylthiophene) on indium tin oxide glass by scanning tunneling microscopy. Chem Mater 5:289–296. doi:10.1021/cm00027a010

127. Kato H, Takemura S, Takakuwa N, Ninomiya K, Watanabe T, Watanabe Y, Nanba N, Hiramatsu T (2006) X-ray photoemission spectroscopy characterization of electrochemical growth of conducting polymer on oxidized Si surface. J Vac Sci Technol A 24:1505–1508. doi:10.1116/1.2208995

128. Kato H, Takemura S (2007) Dynamic force microscopy and X-ray photoemission spectroscopy studies of conducting polymer thin film on nanoscale structured Al surface. Paper presented at the International Society for Optical Engineering conference, in the proceedings 6645:Z, San Diego, CA, 27–30 August 2007. http://spie.org/Publications/Proceedings/Paper/10.1117/12.733602. Accessed 5 Dec 2016

129. Glidle A, Hadyoon CS, Gadegaard N, Cooper JM, Hillman AR, Wilson AR, Ryder KS, Webster JRP, Cubitt R (2005) Evaluating the influence of deposition conditions on solvation of reactive conducting polymers with neutron reflectivity. J Phys Chem B 109: 14335–14343. doi:10.1021/jp0515030

130. Bergqvist J, Mauger S, Tvingstedt K, Arwin H, Inganas O (2013) In situ reflectance imaging of organic thin film formation from solution deposition. Solar Energy Mater Solar Cells 114:89–98. doi:10.1016/j.solmat.2013.02.030

131. Joseph N, Varghese J, Sebastian MT (2015) Self assembled polyaniline nanofibers with enhanced electromagnetic shielding properties. RSC Adv 5:20459–20466. doi:10.1039/c5ra02113h

132. Joo J, Epstein AJ (1994) Electromagnetic-radiation shielding by intrinsically conducting polymers. Appl Phys Lett 65:2278–2280. doi:10.1063/1.112717

133. Courric S, Tran VH (1998) The electromagnetic properties of poly(p-phenylene-vinylene) derivatives. Polymer 39:2399–2408. doi:10.1016/S0032-3861(97)00576-4

134. Masuda K, Tohya HA, Satoh M (2002) Shield strip type low impedance line component using a conducting polymer for a wide frequency band de-coupler set. IEICE Trans Electron E85C:1317–1322. http://apps.webofknowledge.com/full_record.do?product=WOS &search_mode=GeneralSearch&qid=157&SID=Q1qEYwxHJ133rClpT9s&page=1&doc=1. Accessed 5 Dec 2016

135. Ohlan A, Singh K, Chandra A, Dhawan SK (2008) Microwave absorption properties of conducting polymer composite with barium ferrite nanoparticles in 12.4–18 GHz. Appl Phys Lett 93:053114

136. Singh K, Ohlan A, Kotnala RK, Bakhshi AK, Dhawan SK (2008) Dielectric and magnetic properties of conducting ferromagnetic composite of polyaniline with gamma-Fe_2O_3 nanoparticles. Mater Chem Phys 112:651–658. doi:10.1016/j.matchemphys.2008.06.026

137. Phang SW, Tadokoro M, Watanabe J, Kuramoto N (2009) Effect of Fe_3O_4 and TiO_2 addition on the microwave absorption property of polyaniline micro/nanocomposites. Polym Adv Technol 20:550–557. doi:10.1002/pat.1381

138. Dhawan SK, Singh K, Bakhshi AK, Ohlan A (2009) Conducting polymer embedded with nanoferrite and titanium dioxide nanoparticles for microwave absorption. Synth Met 159:2259–2262. doi:10.1016/j.synthmet.2009.08.031

139. Ohlan A, Singh K, Chandra A, Dhawan SK (2010) Shielding and dielectric properties of sulfonic acid-doped pi-conjugated polymer in 8.2–12.4 GHz frequency range. J Appl Polym Sci 115:498–503. doi:10.1002/app.30806

140. Saini P, Arora M, Gupta G, Gupta BK, Singh VN, Choudhary V (2013) High permittivity polyaniline-barium titanate nanocomposites with excellent electromagnetic interference shielding response. Nanoscale 5:4330–4336. doi:10.1039/c3nr00634d

141. Saini P, Arora M (2013) Formation mechanism, electronic properties & microwave shielding by nano-structured polyanilines prepared by template free route using surfactant dopants. J Mater Chem A 1:8926–8934. doi:10.1039/c3ta11086a
142. Faisal M, Khasim S (2013) Polyaniline-antimony oxide composites for effective broadband EMI shielding. Iran Polym J 22:473–480. doi:10.1007/s13726-013-0149-z
143. Faisal M, Khasim S (2013) Broadband electromagnetic shielding and dielectric properties of polyaniline-stannous oxide composites. J Mater Sci-Mater Electron 24:2202–2210
144. Sarvi A, Sundararaj U (2014) Electrical permittivity and electrical conductivity of multi wall carbon nanotube-polyaniline (MWCNT-PANi) core-shell nanofibers and MWCNT-PANi/polystyrene composites. Macromol Mater Eng 299:013–1020. doi:10.1002/mame.201300406
145. Soares BG, Pontes K, Marins JA, Calheiros LF, Livi S, Barra GMO (2015) Poly(vinylidene fluoride-co-hexafluoropropylene)/polyaniline blends assisted by phosphonium—based ionic liquid: dielectric properties and beta-phase formation. Eur Polym J 73:65–74. doi:10.1016/j.eurpolymj.2015.10.003

Chapter 5
Outlook

As it has been presented, in situ electrochemical techniques have been successfully applied in studying the details of both the electrodeposition and the redox transformation of conducting polymers. An even more adequate picture could be gained by hyphenating two in situ electrochemical methods. Every combination proved to be a powerful tool in elucidating the details of the processes both at the molecular and the macroscopic levels.

The investigations brought to sunlight many aspects of the electrochemical processes, by uncovering consequences induced by the electrochemical perturbation. We have known that a good couple of modifications in the properties of the materials are triggered by the charge transfer, and that the achievement of the conducting state is connected with several different alterations.

In the future as a key-question, we should learn a lot about the reasons, why the development of the conducting state is generally delayed compared to the electron transfer, whether it is caused by one or the sum of effects, connected to slower ion, solvent movements, or structural changes induced by these factors, or even just by elongation or shrinking of the polymeric backbone. This problem discussed in Sect. 4.9. could be resolved by applying in situ conductivity and a mass exchange sensitive (EQCN, laser beam deflection, or radiotracer) method in the hyphenated mode, answering, e.g., the question whether neutral species are moving during the pre-polarization or only morphological changes take place to delay the development of the conducting state.

We should go inside into the molecular details of the n-doping processes, similarly to the case of p-doping. In this direction new opportunities are offered during the last years by conducting polymers from ambipolar monomers (cf. 4.10).

Since conducting polymers have already proved their importance in novel, practical applications, the information summarized in the previous chapters is of the highest importance from practical aspect: in order to construct useful devices based on conducting polymers, the knowledge of these details is essential and inevitable.

© The Author(s) 2017

C. Visy, *In situ Combined Electrochemical Techniques for Conducting Polymers*,

SpringerBriefs in Applied Sciences and Technology,

DOI 10.1007/978-3-319-53515-9_5

Moreover, results from the last decade evidenced that combining CPs with in-built components may lead to new materials of synergistic effects. These composites or hybrids may possess properties enabling to elaborate and construct devices which can offer new opportunities in either electrocatalysis, either sensor or photoelectrochemical applications, including special instruments exploiting renewable energy sources.

To consider the past, present, and future of in situ spectroelectrochemistry, a review on the recent state of modern spectroelectrochemistry and trends in the development of spectroelectrochemical techniques has appeared for the combined application of different in situ spectroelectrochemical methods like ESR spectroelectrochemistry, NMR spectroelectrochemistry, Raman spectroelectrochemistry, or IR spectroelectrochemistry to electrode systems [1].

The main part of this review is focused on the advantages of the combined application of spectroelectrochemical techniques in the analysis of electrode reactions. The spectroelectrochemical methods are demonstrated to be generally successful in electrode reactions both for solid structures—like polymers or carbon nanotubes—and molecular structures—like fullerenes, oligothiophenes. Moreover, further progress in spectroscopic techniques—resulting in a variety of triple methods like in situ ESR UV–vis NIR spectroelectrochemistry to make a more detailed characterisation of electrode reaction mechanisms possible—was foreseen.

The most important topics of the rapidly developing field of conducting polymers have been surveyed with emphasis on the problems of synthesis, structure, thermodynamics, and kinetic behavior of these systems. Abundant examples of the growing applications have been also discussed [2].

For the concepts, doubts, proves and still open questions, a comprehensive review has been also published [3].

A final lesson of the above-presented chapters is that all these techniques are not restricted to studies on conducting polymers, but—as it has been illustrated also by examples throughout the book—their use can be extended for all kinds of electroactive layers [4].

Numerous classes of materials can be charged and discharged electronically from external electric circuit or by solute redox agent. The transferred electronic charge has to be compensated by ion exchange between this electroactive layer and the electrolyte solution. Such processes take place in diverse materials as redox or electronically conducting polymers, inorganic solids with mixed-valence transition metals and lithium cation intercalation layers, as well as membranes or porous solids containing a dispersed redox-active component.

So as to follow the development of the field of redox-active surface layers, the attention is called to the scope of the series of the Workshop on the Electrochemistry of Electroactive Materials (WEEM) conference, where contributions cover various aspects of processes that are of importance in the principal classes of modern electroactive materials: electron-conducting polymers, redox-active inorganic materials as well as their hybrid/composite and nanostructured materials [5].

References

1. Dunsch L (2011) Recent advances in in situ multi-spectroelectrochemistry. J Solid State Electrochem 15:1631–1646. doi:10.1007/s10008-011-1453-1
2. Inzelt G, Pineri M, Schultze JW, Vorotyntsev MA (2000) Electron and proton conducting polymers: recent developments and prospects. Electrochim Acta 45:2403–2421. doi:10.1016/S0013-4686(00)00329-7
3. Heinze J, Frontana-Uribe BA, Ludwigs S (2010) Electrochemistry of conducting polymers; persistent models and new concepts. Chem Rev 110:4724–4771. doi:10.1021/cr900226k
4. Hillman AR, Pickup P, Seeber R, Skompska M, Vorotyntsev MA (2014) Electrochemistry of electroactive materials foreword. Electrochim Acta 122:1–2. doi:10.1016/j.electacta.2014.01.001
5. Vorotyntsev AM, Scholz F (2007) Special issue with contributions to the conference "International Workshop on Electrochemistry of Electroactive Materials" (WEEM-2006), Repino, St-Petersburg Region, Russia, 23–28 June 2006. J Solid State Electrochem 11:1007. doi:10.1007/s10008-007-0297-1

Erratum to: Applications of the in situ Combined Electrochemical Techniques: Problems and Answers Attempted by the in situ Combined Methods

Csaba Visy

Erratum to:
Chapter 4 in: C. Visy, *In situ Combined*
***Electrochemical Techniques for Conducting Polymers*,**
SpringerBriefs in Applied Sciences and Technology,
DOI 10.1007/978-3-319-53515-9_4

In the original version of the book, in Chapter 4, page 49, the typographical error "D" in the chemical equation "$P^+ + P^+ D \rightleftarrows PP^{++}$" should be deleted. The erratum chapter and the book have been updated with the change.

The updated original online version for this chapter can be found at
http://dx.doi.org/10.1007/978-3-319-53515-9_4

Index

A
Absorbance, 26, 30, 43
Absorption, 9, 19, 25, 33, 42, 43, 46, 47
A.c. impedance, 36
Ambipolar, 43, 63
Atomic force microscpy (AFM), 14, 19

C
Chronovoltabsorptometric, 50
Components of the total charge, 36
Conductance, 12, 15, 26, 29, 30, 38, 39, 41,
    46–48
Conductivity, 12, 14, 28, 30, 38, 41, 43, 45, 50

E
EIS, 18, 52
Electrochemical impedance, 11, 12, 18, 20
Electrogravimetry, 16–18, 32
Electromagnetic shielding, 53
Electron spin resonance (ESR), 9, 14, 15, 25,
    36, 40, 42, 45–48, 51
Ellipsometric/Ellipsometry, 13, 16, 18, 32, 34,
    37, 44, 52
EPR, 40, 44, 46
EQCM, 10, 26, 27, 32, 33, 38, 52
EQCN, 10, 29, 31
ESR/UV-vis-NIR, 48

F
Factor analysis, 8, 44
Fourier Transform Infra Red
    spectroscopy (FTIR), 8, 16, 18, 20, 26, 34,
    35, 41, 44, 48, 50–52

G
Gravimetric, 32

H
Hysteresis, 38

I
Infrared, 26, 34
In situ conductivity, 35, 36, 43, 45
IR, 9, 16, 25, 26, 30, 35

L
Laser beam deflection, 14, 29

M
Mirage effect, 35

N
NMR, 10

O
Open circuit relaxation, 29
Optical, 37, 42, 46

P
PDB, 17
Phase transition, 34

R
Radiotracer, 11
Raman, 8, 35, 42–46
Reflectance, 16, 19, 50, 52, 53

S
Scanning Electrochemical Microscopy
    (SECM), 17, 19, 26
Scanning Tunneling Microscopy, 52
Shrinking, 37
Spectral, 37, 41

© The Author(s) 2017
C. Visy, *In situ Combined Electrochemical Techniques for Conducting Polymers*,
SpringerBriefs in Applied Sciences and Technology,
DOI 10.1007/978-3-319-53515-9